Advanced C and C++ Compiling

Milan Stevanovic

Apress·

Advanced C and C++ Compiling

ISBN-13 (pbk): 978-1-4302-6667-9

ISBN-13 (electronic): 978-1-4302-6668-6

President and Publisher: Paul Manning
Lead Editor: Michelle Lowman
Developmental Editor: James Markham
Technical Reviewers: Nemanja Trifunović, Ben Combee, Miroslav Ristić
Editorial Board: Steve Anglin, Mark Beckner, Ewan Buckingham, Gary Cornell, Louise Corrigan, Jim DeWolf, Jonathan Gennick, Jonathan Hassell, Robert Hutchinson, Michelle Lowman, James Markham, Matthew Moodie, Jeff Olson, Jeffrey Pepper, Douglas Pundick, Ben Renow-Clarke, Dominic Shakeshaft, Gwenan Spearing, Matt Wade, Steve Weiss
Coordinating Editor: Jill Balzano
Copy Editor: Mary Behr
Compositor: SPi Global
Indexer: SPi Global
Artist: SPi Global
Cover Designer: Anna Ishchenko

Distributed to the book trade worldwide by Springer Science+Business Media New York, 233 Spring Street, 6th Floor, New York, NY 10013. Phone 1-800-SPRINGER, fax (201) 348-4505, e-mail orders-ny@springer-sbm.com, or visit www.springeronline.com. Apress Media, LLC is a California LLC and the sole member (owner) is Springer Science + Business Media Finance Inc (SSBM Finance Inc). SSBM Finance Inc is a Delaware corporation.

For information on translations, please e-mail rights@apress.com, or visit www.apress.com.

Apress and friends of ED books may be purchased in bulk for academic, corporate, or promotional use. eBook versions and licenses are also available for most titles. For more information, reference our Special Bulk Sales–eBook Licensing web page at www.apress.com/bulk-sales.

Any source code or other supplementary material referenced by the author in this text is available to readers at www.apress.com. For detailed information about how to locate your book's source code, go to www.apress.com/source-code/.

The book is dedicated to Milena, Pavle, and Selina.

Contents at a Glance

Contents

About the Author

Milan Stevanović is a senior multimedia software consultant based in the San Francisco Bay Area. The span of his engineering experience covers a multitude of disciplines, ranging from the board level analog and digital hardware design, and assembly programming, all the way to C/C++ design and software architecture. His professional focus has been in the domain of the analysis of a variety of compressed multimedia formats, and design related to a variety of multimedia frameworks (GStreamer, DirectX, OpenMAX, ffmpeg) in Linux (desktop, embedded, Android native) as well as in Windows.

He has designed multimedia software for numerous companies (C-Cube, Philips, Harman, Thomson, Gracenote, Palm, Logitech, Panasonic, Netflix), creating a variety of cutting edge technology products (ZiVA-1 and ZiVA-3 DVD decoder chip, Philips TriMedia media processor, Palm Treo and Palm Pre mobile phones, Netflix application for Android).

He is an original member of the developer duo (together with David Ronca) that created the avxsynth open source project (the first successful port of avisynth to Linux).

He holds a MSEE degree from Purdue University (1994) and an undergraduate degree in EE (1987) and Music – flute performance (1990) from the University of Belgrade.

About the Technical Reviewers

Nemanja Trifunović was born in Kragujevac, Serbia. Now he lives in Boston area with his wife and daughters. He wrote his first program at the age of 13 on a Sinclair Spectrum and became a professional software developer after he graduated. Currently he is a Development Expert at SAP, working on HANA database system. In the past he worked at Microsoft, Lionbridge and Lernout & Hauspie Speech Products. He is one of the most distinguished Code Project contributors of programming-related articles (see www.codeproject.com/Members/Nemanja-Trifunovic, www.codeproject.com/Articles/Nemanja-Trifunovic#articles).

Ben Combee is a lead developer on the Enyo JS framework. In a past life, he worked on the CodeWarrior for Palm OS and CodeWarrior for Win32 tools at Metrowerks, plus many projects at Palm, including the Foleo netbook and on the architecture of Palm webOS.

Miroslav Ristić, by a chain of events, instead of becoming a jet fighter pilot, graduated and received his MSc in Computer Engineering at University of Novi Sad, Serbia. Currently he lives in the San Francisco Bay area, pursuing his career goals. His interests span a wide variety of topics. He is very passionate about creating things from scratch. He is happily married to the extraordinary and astounding lady of his dreams, Iva.

Acknowledgments

There are many people who have made a lasting impact on the way I think and reason, and how I see the world, especially the world of technology. Since this is my first published book, which I imagine is a special occasion for every author, I'll take the freedom to express my gratitude to a long list of people.

Had I not encountered a collection of superstar professors teaching at my 12th Belgrade Gimnasium, my entire life path would have probably taken a significantly different direction (no, I wouldn't have ended up on the wrong side of the law; I would have probably become a professional musician/ arranger/composer, a Quincy Jones or Dave Grusin of a kind). The math skills I gained as a student of professor Stevan Šijački were my ticket to EE undergraduate studies, even though a music career was originally planned. My physics professor Ljubinka Prokić and her insisting on clarity and precision, plus her love and devotion to both physics and to her students, shaped my approach toward knowledge in general, summed up best by Einstein's "if you can't explain it simply, you don't understand it well enough." Finally, the language and literature professor Jelena Hristodulo taught me the most important of all lessons: that those who dive deeper for "whys" instead of swimming on the surface between "whats" are bound to truly understand the world around them and substantially change it. Even though she had poets and philosophers in mind, I find that this recipe applies surprisingly well to people with technical backgrounds.

Without Dr. George Wodicka, whose research assistant I was at Purdue University, I probably wouldn't be in the US today, so 20+ years of my Silicon Valley career wouldn't have happened, and very likely this book would have never been written.

The encouragement from Dr. George Adams of Purdue University to take his Computational Models and Methods core course was a decisive first step on the long journey of my career transformation from a young hardware design engineer to a seasoned software professional with 20+ years of experience.

The strong belief of David Berkowitz, my manager at Palm, that my multimedia design skills could and should expand into Linux territory was one of the decisive moments of my career. His unique people skills and his ability to create a cohesive team made my time in his Palm Multimedia Group team a memorable experience. The atmosphere in the Palm cafeteria after watching the video broadcast of the Palm Pre presentation that blew the tech world out of its socks at the CES show in Las Vegas made that January 08, 2009 the single most memorable day of my professional career. My engineering skillset became enriched by plenty of significant experiences and directions, some of which directly led me to the material presented in this book.

Working with Saldy Antony was another truly inspirational experience. After the years we spent together in the Philips TriMedia team, our careers went in different directions. While I stayed deeply immersed in the immediate details of multimedia, he tremendously spread his skillsets into the domain of software architecture and software management. When our paths crossed again at 2-Wire and later at Netflix, the open source era has already stepped into the life of a professional multimedia development. With each passing day, the everyday work of the multimedia software professional meant a lot less writing of code and more integrating existing third party/open source code. The talks with Saldy in which he tried to convince me that I should enhance my skillset with the skills of a software build engineer definitely had some effect on me. However, seeing him in action, rolling up his sleeves in the spare time between the management-level meetings, joining the team in the cubicles and swiftly resolving any problem related to the compiler, linker, libraries, and code deployment issues definitely had a lasting impression on me. This was when I decided to "hug the monster" and learn what I originally did not consider vital to my personal role in the industry.

The invitation from David Ronca, the manager of the Netflix Encoding Technology Group, to work on something really interesting is another cornerstone in the journey of creating this book. The "something really interesting" was the open source project of converting the popular avisynth Windows video post-production tool to Linux, the project known as *avxsynth*. His exceptionally clear vision and firmly set architectural requirements combined with the freedom he gave me to investigate the implementation details ultimately led to tremendous success. The project, accomplished in a period of only 2.5 months, was also an immense learning experience for me. Encountering and surpassing the difficulties along the way required spending hours researching topics and forming my personal treasury of related tips and tricks. Daily talks with my group members (Dr. Anne Aaron, Pradip Gajjar, and especially Brian Feinberg, the walking archive of rare pieces of knowledge) over David's signature lattes have helped me get an idea how much more I still have to learn.

The encounter with Apress Editorial Director Steve Anglin was a movie-like experience. The last time I saw anyone even remotely similar to him was when I was watching a detective TV series in 1970s as a kid back home in Belgrade, Serbia. Smart, communicative, quick thinking, immediately recognizing the right things, reacting on a hunch in the nick of time, and straight to the point, he was the kind of professional I almost stopped believing ever existed. Collaboration with him made the process of publishing this book a memorable experience.

The Apress Acquisition Editor Michelle Lowman carried through the decisive effort to proofread and present the book materials through the several rounds of Apress team discussions, for which I am deeply thankful.

My special thanks also goes to my Apress editorial team (Ewan Buckingham, Kevin Shea, Jill Balzano, James Markham, and the army of others with whom I haven't had direct contact). The 'british lord' character which I use in the book to point out the nature of the executable file was not in fact inspired by Ewan Buckingham, but it would not be a mistake if it were. His control over the major flow of the publishing effort was sovereign and right to the point in many occasions.

Without the last minute intervention from my talented niece Jovana Stefanovic and her Brosnan-Clooney DSP algorithm, my cover photo would look like just like me, or maybe even worse.

Finally, the input from my team of proofreaders (Nemanja Trifunović, Miroslav Ristić, Ben Combee), support from friends (David Moffat, Pradip Gajjar) and from the mysterious bunch collectively known as "Arques 811 ex-Philips group" (especially Daniel Ash and Thierry Seegers) has proven to be very valuable. I am deeply thankful for their efforts and time taken from their busy professional lives to provide feedback about the book's content.

Finally, without the love and support from my wife, Milena, son, Pavle, and daughter, Selina, and their patience during many weekends and evenings spent working on the book's material, the whole project would not have happened.

Introduction

It took me quite some time to become aware of an amazing analogy that exists between the culinary art and the art of computer programming.

Probably the most obvious comparison that comes to mind is that both the culinary specialist and the programmer have similar ultimate goals: to feed. For a chef, it is the human being, for which plenty of raw ingredients are used to provide edible nutrients as well as gastronomic pleasure, whereas for the programmer it is the microprocessor, for which a number of different procedures are used to provide the code that not only needs to produce some meaningful actions, but also needs to be delivered in the optimum form.

As much as this introductory comparison point may seem a bit far-fetched or even childish, the subsequent comparison points are something that I find far more applicable and far more convincing.

The recipes and instructions for preparing dishes of all kinds are abundant and ubiquitous. Almost every popular magazine has a culinary section dedicated to all kinds of foods, and all kind of food preparation scenarios, ranging from quick-and-easy/last-minute recipes all the way to really elaborate ones, from ones focusing on nutrition tables of ingredients to ones focusing on the delicate interplay between extraordinary, hard-to-find ingredients.

However, at the next level of expertise in the culinary art, the availability of resources drops exponentially. The recipes and instructions for running the food business (volume production, running the restaurant, or catering business), planning the quantities and rhythm of delivery for food preparation process, techniques and strategies for optimizing the efficiency of food delivery, techniques for choosing the right ingredients, minimizing the decay of stored ingredients—this kind of information is substantially more hard to find. Rightfully so, as these kinds of topics delineate the difference between amateur cooking and the professional food business.

The situation with programming is quite similar.

The information about a vast variety of programming languages is readily available, through thousands of books, magazines, articles, web forums, and blogs, ranging from the absolute beginner level all the way to the "prepare for the Google programming interview" tips.

These kinds of topics, however, cover only about half of the skills required by the software professional. Soon after the immediate gratification of seeing the program we created actually executing (and doing it right) comes the next level of important questions: how to architect the code to allow for easy further modifications, how to extract reusable parts of the functionality for future use, how to allow smooth adjustment for different environments (starting from different human languages and alphabets, all the way to running on different operating systems).

As compared to the other topics of programming, these kinds of topics are rarely discussed, and to this day belong to the form of "black art" reserved for a few rare specimens of computer science professionals (mostly software architects and build engineers) as well as to the domain of university-level classes related to the compiler/linker design.

One particular factor—the ascent of Linux and the proliferation of its programming practices into a multitude of design environments—has brought a huge impetus for a programmer to pay attention to these topics. Unlike the colleagues writing software for "well-cushioned" platforms (Windows and Mac, in which the platform, IDEs, and SDKs relieve the programmer of thinking about certain programming aspects), a Linux programmer's daily routine is to combine together the code coming from variety of sources, coding practices, and in forms which require immediate understanding of inner workings of compiler, linker, the mechanism of program loading, and hence the details of designing and using the various flavors of libraries.

The purpose of this book is to discuss a variety of valuable pieces of information gathered from a scarce and scattered knowledge base and validate it through a number of carefully tailored simple experiments. It might be important to point out that the author does not come from a computer science background. His education on the topic came as a result of being immersed as electrical engineer in the technology frontier of the Silicon Valley multimedia industry in the time of the digital revolution, from the late 90s to the present day. Hopefully, this collection of topics will be found useful by a wider audience.

Audience (Who Needs This Book and Why)

The side effect of myself being a (very busy, I must say proudly) software design hands-on consultant is that I regularly come in contact with an extraordinary variety of professional profiles, maturity, and accomplishment levels. The solid statistic sample of the programmer population (of Silicon Valley, mostly) that I meet by switching office environments several times during a work week has helped me get a fairly good insight into the profiles of who may benefit from reading this book. So, here they are.

The first group is made of the C/C++ programmers coming from a variety of engineering backgrounds (EE, mechanical, robotics and system control, aerospace, physics, chemistry, etc.) who deal with programming on a daily basis. A lack of formal and more focused computer science education as well as a lack of non-theoretical literature on the topic makes this book a precious resource for this particular group.

The second group is comprised of junior level programmers with a computer science background. This book may help concretize the body of their existing knowledge gained in core courses and focus it to the operational level. Keeping the quick summaries of Chapters 12–14 somewhere handy may be worthwhile even for the more senior profiles of this particular group.

The third group is made of folks whose interest is in the domain of OS integration and customization. Understanding the world of binaries and the details of their inner working may help "clean the air" tremendously.

About the Book

Originally, I did not have any plans to write this particular book. Not even a book in the domain of computer science. (Signal processing? Art of programming? Maybe...but a computer science book? Naaah...)

The sole reason why this book emerged is the fact that through the course of my professional career I had to deal with certain issues, which at that time I thought someone else should take care of.

Once upon a time, I made the choice of following the professional path of a high-tech assassin of sort, the guy who is called by the citizens of the calm and decent high tech communities to relieve them from the terror of nasty oncoming multimedia-related design issues wreaking havoc together with a gang of horrible bugs. Such a career choice left pretty much no space for exclusivity in personal preferences typically found by the kids who would eat the chicken but not the peas. The ominous "or else" is kind of always there. Even though FFTs, wavelets, Z-transform, FIR and IIR filters, octaves, semitones, interpolations and decimations are naturally my preferred choice of tasks (together with a decent amount of C/C++ programming), I had to deal with issues that would not have been my personal preference. Someone had to do it.

Surprisingly, when looking for the direct answers to very simple and pointed questions, all I could find was a scattered variety of web articles, mostly about the high-level details. I was patiently collecting the "pieces of the puzzle" together, and managed to not only complete the design tasks at hand but also to learn along the way.

One fine day, the time came for me to consolidate my design notes (something that I regularly do for the variety of topics I deal with). This time, however, when the effort was completed, it all looked...well...like a book. This book. Anyways...

Given the current state of the job market, I am deeply convinced that (since about the middle of the first decade of 21st century) knowing the C/C++ language intricacies perfectly—and even algorithms, data structures, and design patterns—is simply not enough.

In the era of open source, the life reality of the professional programmer becomes less and less about "knowing how to write the program" and instead substantially more about "knowing how to integrate existing bodies of code." This assumes not only being able to read someone else's code (written in variety of coding styles and practices), but also knowing the best way to integrate the code with the existing packages that are mostly available in binary form (libraries) accompanied by the export header files.

Hopefully, this book will both educate (those who may need it) as well as provide the quick reference for the most of the tasks related to the analysis of the C/C++ binaries.

Why am I illustrating the concepts mostly in Linux?

It's nothing personal.

In fact, those who know me know how much (back in the days when it was my preferred design platform) I used to like and respect the Windows design environment—the fact that it was well documented, well supported, and the extent to which the certified components worked according to the specification. A number of professional level applications I've designed (GraphEdit for Windows Mobile for Palm, Inc., designed from scratch and crammed with extra features being probably the most complex one, followed by a number of media format/DSP analysis applications) has led me toward the thorough understanding and ultimately respect for the Windows technology at the time.

In the meantime, the Linux era has come, and that's a fact of life. Linux is everywhere, and there is little chance that a programmer will be able to ignore and avoid it.

The Linux software design environment has proven itself to be open, transparent, simple and straight to-the-point. The control over individual programming stages, the availability of well-written documentation, and even more "live tongues" on the Web makes working with the GNU toolchain a pleasure.

The fact that the Linux C/C++ programming experience is directly applicable to low-level programming on MacOS contributed to the final decision of choosing the Linux/GNU as the primary design environment covered by this book.

But, wait! Linux and GNU are not exactly the same thing!!!

Yes, I know. Linux is a kernel, whereas GNU covers whole lot of things above it. Despite the fact that the GNU compiler may be used on the other operating systems (e.g. MinGW on Windows), for the most part the GNU and Linux go hand-in-hand together. To simplify the whole story and come closer to how the average programmer perceives the programming scene, and especially in contrast with the Windows side, I'll collectively refer to GNU + Linux as simply "Linux."

The Book Overview

Chapters 2–5 are mostly preparing the terrain for making the point later on. The folks with the formal computer science background probably do not need to read these chapters with focused attention (fortunately, these chapters are not that long). In fact, any decent computer science textbook may provide the same framework in far more detail. My personal favorite is Bryant and O'Hallaron's *Computer Systems - A Programmer's Perspective* book, which I highly recommend as a source of nicely arranged information related to the broader subject.

Chapters 6–12 provide the essential insight into the topic. I invested a lot of effort into being concise and trying to combine words and images of familiar real-life objects to explain the most vital concepts whose understanding is a must. For those without a formal computer science background, reading and understanding these chapters is highly recommended. In fact, these chapters represent the gist of the whole story.

Chapters 13–15 are kind of a practical cheat sheet, a form of neat quick reminders. The platform-specific set of tools for the binary files analyses are discussed, followed by the cross-referencing "How Tos" part which contains quick recipes of how to accomplish certain isolated tasks.

Appendix A contains the technical details of the concepts mentioned in Chapter 8. Appendix A is available online only at www.apress.com. For detailed information about how to locate it, go to www.apress.com/source-code/. After understanding the concepts from Chapter 8, it may be very useful to try to follow the hands-on explanations of how and why certain things really work. I hope that a little exercise may serve as practical training for the avid reader.

CHAPTER 1

■ ■ ■

Multitasking OS Basics

The ultimate goal of all the art related to building executables is to establish as much control as possible over the process of program execution. In order to truly understand the purpose and meaning of certain parts of the executable structure, it is of the utmost importance to gain the full understanding of what happens during the execution of a program, as the interplay between the operating system kernel and the information embedded inside the executable play the most significant roles. This is particularly true in the initial phases of execution, when it is too early for runtime impacts (such as user settings, various runtime events, etc.) which normally happen.

The mandatory first step in this direction is to understand the surroundings in which the programs operate. The purpose of this chapter is to provide in broad sketches the most potent details of a modern multitasking operating system's functionality.

Modern multitasking operating systems are in many aspects very close to each other in terms of how the most important functionality is implemented. As a result, a conscious effort will be made to illustrate the concepts in platform-independent ways first. Additionally, attention will be paid to the intricacies of platform-specific solutions (ubiquitous Linux and ELF format vs. Windows) and these will be analyzed in great detail.

Useful Abstractions

Changes in the domain of computing technology tend to happen at very fast pace. The integrated circuits technology delivers components that are not only rich in variety (optical, magnetic, semiconductor) but are also getting continually upgraded in terms of capabilities. According to the Moore's Law, the number of transistors on integrated circuits doubles approximately every two years. Processing power, which is tightly associated with the number of available transistors, tends to follow a similar trend.

As was found out very early on, the only way of substantially adapting to the pace of change is to define overall goals and architecture of computer systems in an abstract/generalized way, at the level above the particulars of the ever-changing implementations. The crucial part of this effort is to formulate the abstraction in such a way that any new actual implementations fit in with the essential definition, leaving aside the actual implementation details as relatively unimportant. The overall computer architecture can be represented as a structured set of abstractions, as shown in Figure 1-1.

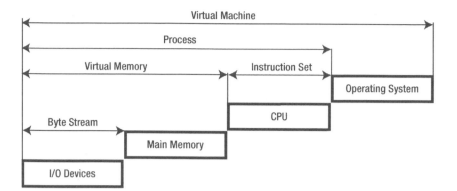

Figure 1-1. *Computer Architecture Abstractions*

The abstraction at the lowest level copes with the vast variety of I/O devices (mouse, keyboard, joystick, trackball, light pen, scanner, bar code readers, printer, plotter, digital camera, web camera) by representing them with their quintessential property of byte stream. Indeed, regardless of the differences between various devices' purposes, implementations, and capabilities, it is the byte streams these devices produce or receive (or both) that are the detail of utmost importance from the standpoint of computer system design.

The next level abstraction, the concept of virtual memory, which represents the wide variety of memory resources typically found in the system, is the subject of extraordinary importance for the major topic of this book. The way this particular abstraction actually represents the variety of physical memory devices not only impacts the design of the actual hardware and software but also lays a groundwork that the design of compiler, linker, and loader relies upon.

The instruction set that abstracts the physical CPU is the abstraction of the next level. Understanding the instruction set features and the promise of the processing power it carries is definitely the topic of interest for the master programmer. From the standpoint of our major topic, this level of abstraction is not of primary importance and will not be discussed in great detail.

The intricacies of the operating system represent the final level of abstraction. Certain aspects of the operating system design (most notably, multitasking) have a decisive impact on the software architecture in general. The scenarios in which the multiple parties try to access the shared resource require thoughtful implementation in which unnecessary code duplication would be avoided—the factor that directly led to the design of shared libraries.

Let's make a short detour in our journey of analyzing the intricacies of the overall computer system and instead pay special attention to the important issues related to memory usage.

Memory Hierarchy and Caching Strategy

There are several interesting facts of life related to the memory in computer systems:

- The need for memory seems to be insatiable. There is always a need for far more than is currently available. Every quantum leap in providing larger amounts (of faster memory) is immediately met with the long-awaiting demand from the technologies that have been conceptually ready for quite some time, and whose realization was delayed until the day when physical memory became available in sufficient quantities.

- The technology seems to be far more efficient in overcoming the performance barriers of processors than memory. This phenomenon is typically referred to as "the processor-memory gap."

- The memory's access speed is inversely proportional to the storage capacity. The access times of the largest capacity storage devices are typically several orders of magnitude larger than that of the smallest capacity memory devices.

Now, let's take a quick look at the system from the programmer/designer/engineer point of view. Ideally, the system needs to access all the available memory as fast as possible—which we know is never possible to achieve. The immediate next question then becomes: is there anything we can do about it?

The detail that brings tremendous relief is the fact that the system does not use all the memory all of the time, but only some memory for some of the time. In that case, all we really need to do is to reserve the fastest memory for running the immediate execution, and to use the slower memory devices for the code/data that is not immediately executed. While the CPU fetches from the fast memory the instructions scheduled for the immediate execution, the hardware tries to guess which part of the program will be executed next and supplies that part of the code to the slower memory to await the execution. Shortly before the time comes to execute the instructions stored in the slower memory, they got transferred into the faster memory. This principle is known as caching.

The real-life analogy of caching is something that an average family does with their food supply. Unless we live in very isolated places, we typically do not buy and bring home all the food needed for a whole year. Instead, we mostly maintain moderately large storage at home (fridge, pantry, shelves) in which we keep a food supply sufficient for a week or two. When we notice that these small reserves are about to be depleted, we make a trip to the grocery and buy only as much food as needed to fill up the local storage.

The fact that a program execution is typically impacted by a number of external factors (user settings being just one of these) makes the mechanism of caching a form of guesswork or a hit-or-miss game. The more predictable the program execution flows (measured by the lack of jumps, breaks, etc.) the smoother the caching mechanism works. Conversely, whenever a program encounters the flow change, the instructions that were previously accumulated end up being discarded as no longer needed, and a new, more appropriate part of the program needs to be supplied from the slower memory.

The implementation of a caching principle is omnipresent and stretches across several levels of memory, as illustrated in Figure 1-2.

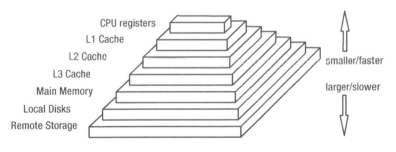

Figure 1-2. *Memory caching hierarchy principle*

Virtual Memory

The generic approach of memory caching gets the actual implementation on the next architectural level, in which the running program is represented by the abstraction called **process**.

Modern multitasking operating systems are designed with the intention to allow one or more users to concurrently run several programs. It is not unusual for the average user to have several applications (e.g. web browser, editor, music player, calendar) running simultaneously.

The disproportion between the needs of the memory and the limited memory availability was resolved by the concept of **virtual memory**, which can be outlined by the following set of guidelines:

- Program memory allowances are fixed, equal for all programs, and declarative in nature.

    Operating systems typically allow the program (process) to use 2^N bytes of memory, where N is nowadays 32 or 64. This value is fixed and is independent of the availability of the physical memory in the system

- The amount of physical memory may vary. Usually, memory is available in quantities that are several times smaller than the declared process address space. It is nothing unusual that the amount of physical memory available for running programs is an uneven number.

- Physical memory at runtime is divided into small fragments (pages), with each page being used for programs running simultaneously.

- The complete memory layout of the running program is kept on the slow memory (hard disk). Only the parts of the memory (code and data) that are about to be currently executed are loaded into the physical memory page.

The actual implementation of the virtual memory concept requires the interaction of numerous system resources such as hardware (hardware exceptions, hardware address translation), hard disk (swap files), as well as the lowest level operating system software (kernel). The concept of virtual memory is illustrated in Figure 1-3.

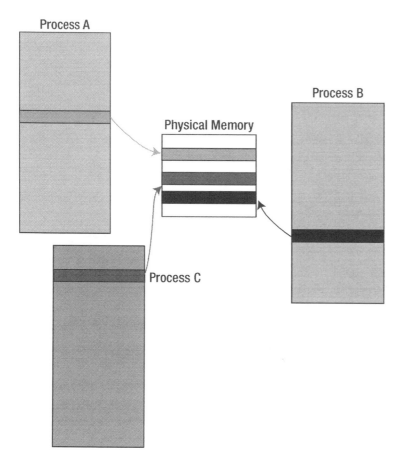

Figure 1-3. *Virtual memory concept implementation*

Virtual Addressing

The concept of virtual addressing is at the very foundation of the virtual memory implementation, and in many ways significantly impacts the design of compilers and linkers.

As a general rule, the program designer is completely relieved of worrying about the addressing range that his program will occupy at runtime (at least this is true for the majority of user space applications; kernel modules are somewhat exceptional in this sense). Instead, the programming model assumes that the address range is between 0 and 2^N (virtual address range) and is the same for all programs.

The decision to grant a simple and unified addressing scheme for all programs has a huge positive impact on the process of code development. The following are the benefits:

- Linking is simplified.

- Loading is simplified.

- Runtime process sharing becomes available.

- Memory allocation is simplified.

The actual runtime placement of the program memory in a concrete address range is performed by the operating system through the mechanism of address translation. Its implementation is performed by the hardware module called a memory management unit (MMU), which does not require any involvement of the program itself.

Figure 1-4 compares the virtual addressing mechanism with a plain and simple physical addressing scheme (used to this day in the domain of simple microcontroller systems).

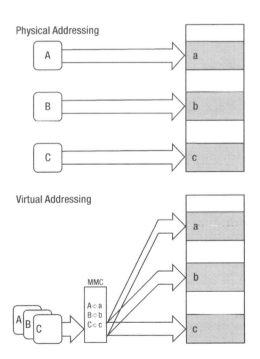

Figure 1-4. *Physical vs. virtual addressing*

Process Memory Division Scheme

The previous section explanted why it is possible to provide the identical memory map to the designer of (almost) any program. The topic of this section is to discuss the details of the internal organization of the process memory map. It is assumed that the program address (as viewed by the programmer) resides in the address span between 0 and 2^N, N being 32 or 64.

Various multitasking/multiuser operating systems specify different memory map layouts. In particular, the Linux process virtual memory map follows the mapping scheme shown in Figure 1-5.

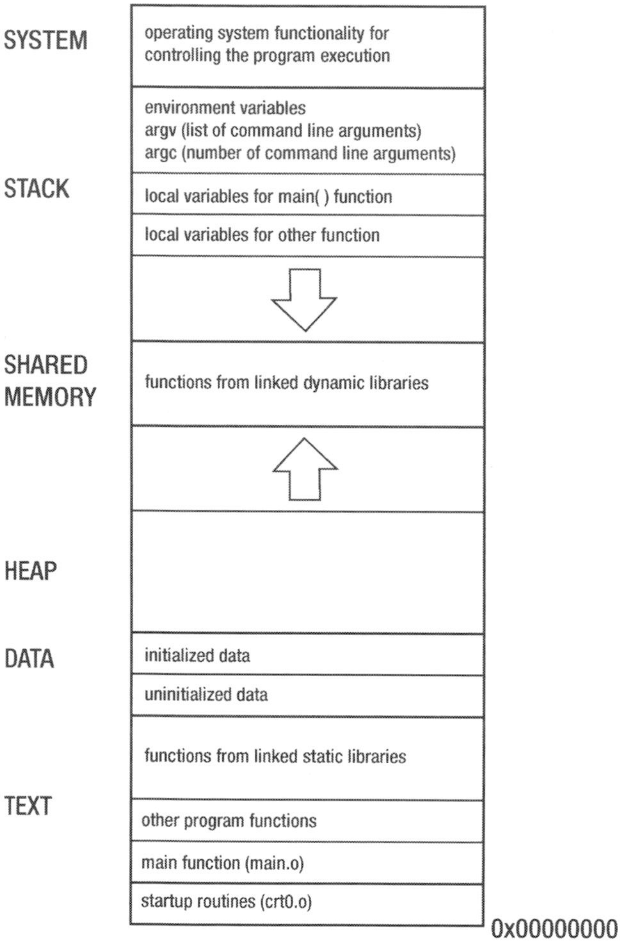

SYSTEM	operating system functionality for controlling the program execution
STACK	environment variables argv (list of command line arguments) argc (number of command line arguments)
	local variables for main( ) function
	local variables for other function
	⬇
SHARED MEMORY	functions from linked dynamic libraries
	⬆
HEAP	
DATA	initialized data
	uninitialized data
TEXT	functions from linked static libraries
	other program functions
	main function (main.o)
	startup routines (crt0.o)

0x00000000

Figure 1-5. *Linux process memory map layout*

Regardless of the peculiarities of a given platform's process memory division scheme, the following sections of the memory map must be always supported:

- **Code section** carrying the machine code instructions for the CPU to execute (**.text** section)

- **Data sections** carrying the data on which the CPU will operate. Typically, separate sections are kept for initialized data (**.data** section), for uninitialized data (**.bss** section), as well as for constant data (**.rdata** section)

- The **heap** on which the dynamic memory allocation is run

- The **stack,** which is used to provide independent space for functions

- The topmost part belonging to the kernel where (among the other things) the process-specific environment variables are stored

A beautifully detailed discussion of this particular topic written by Gustavo Duarte can be found at http://duartes.org/gustavo/blog/post/anatomy-of-a-program-in-memory.

The Roles of Binaries, Compiler, Linker, and Loader

The previous section shed some light on the memory map of the running process. The important question that comes next is how the memory map of the running process gets created at runtime. This section will provide an elementary insight into that particular side of the story.

In a rough sketch,

- The program binaries carry the details of the blueprint of the running process memory map.

- The skeleton of a binary file is created by the linker. In order to complete its task, the linker combines the binary files created by the compiler in order to fill out the variety of memory map sections (code, data, etc.).

- The task of initial creation of the process memory map is performed by the system utility called the program loader. In the simplest sense, the loader opens the binary executable file, reads the information related to the sections, and populates the process memory map structure.

This division of roles pertains to all modern operating systems.

Please be aware that this simplest description is far from providing the whole and complete picture. It should be taken as a mild introduction into the subsequent discussions through which substantially more details about the topic of binaries and process loading will be conveyed as we progress further into the topic.

Summary

This chapter provided an overview of the concepts that most fundamentally impact the design of modern multitasking operating systems. The cornerstone concepts of virtual memory and virtual addressing not only affect the program execution (which will be discussed in detail in the next chapter), but also directly impact how the program executable files are built (which will be explained in detail later in the book).

CHAPTER 2

■ ■ ■

Simple Program Lifetime Stages

In the previous chapter, you obtained a broad insight into aspects of the modern multitasking operating system's functionality that play a role during program execution. The natural next question that comes to the programmer's mind is what to do, how, and why in order to arrange for the program execution to happen.

Much like the lifetime of a butterfly is determined by its caterpillar stage, the lifetime of a program is greatly determined by the inner structure of the binary, which the OS loader loads, unpacks, and puts its contents into the execution. It shouldn't come as a big surprise that most of our subsequent discussions will be devoted to the art of preparing a blueprint and properly embedding it into the body of the binary executable file(s). We will assume that the program is written in C/C++.

To completely understand the whole story, the details of the rest of the program's lifetime, the loading and execution stage, will be analyzed in great detail. Further discussions will be focused around the following stages of the program's lifetime:

1. Creating the source code
2. Compiling
3. Linking
4. Loading
5. Executing

The truth be told, this chapter will cover far more details about the compiling stage than about the subsequent stages. The coverage of subsequent stages (especially linking stage) only starts in this chapter, in which you will only see the proverbial "tip of the iceberg." After the most basic introduction of ideas behind the linking stage, the remainder of the book will deal with the intricacies of linking as well as program loading and executing.

Initial Assumptions

Even though it is very likely that a huge percentage of readers belong to the category of advanced-to-expert programmers, I will start with fairly simple initial examples. The discussions in this chapter will be pertinent to the very simple, yet very illustrative case. The demo project consists of two simple source files, which will be first compiled and then linked together. The code is written with the intention of keeping the complexity of both compiling and linking at the simplest possible level.

In particular, no linking of external libraries, particularly not dynamic linking, will be taking place in this demo example. The only exception will be the linking with the C runtime library (which is anyways required for the vast majority of programs written in C). Being such a common element in the lifetime of C program execution, for the sake of simplicity I will purposefully turn a blind eye to the particular details of linking with the C runtime library, and assume that the program is created in such a way that all the code from the C runtime library is "automagically" inserted into the body of the program memory map.

By following this approach, I will illustrate the details of program building's quintessential problems in a simple and clean form.

Code Writing

Given that the major topic of this book is the process of program building (i.e., what happens after the source code is written), I will not spend too much time on the source code creation process.

Except in a few rare cases when the source code is produced by a script, it is assumed that a user does it by typing in the ASCII characters in his editor of choice in an effort to produce the written statements that satisfy the syntax rules of the programming language of choice (C/C++ in our case). The editor of choice may vary from the simplest possible ASCII text editor all the way to the most advanced IDE tool. Assuming that the average reader of this book is a fairly experienced programmer, there is really nothing much special to say about this stage of program life cycle.

However, there is one particular programming practice that significantly impacts where the story will be going from this point on, and it is worth of paying extra attention to it. In order to better organize the source code, programmers typically follow the practice of keeping the various functional parts of the code in separate files, resulting with the projects generally comprised of many different source and header files.

This programming practice was adopted very early on, since the time of the development environments made for the early microprocessors. Being a very solid design decision, it has been practiced ever since, as it is proven to provide solid organization of the code and makes code maintenance tasks significantly easier.

This undoubtedly useful programming practice has far reaching consequences. As you will see soon, practicing it leads to certain amount of indeterminism in the subsequent stages of the building process, the resolving of which requires some careful thinking.

Concept illustration: Demo Project

In order to better illustrate the intricacies of the compiling process, as well as to provide the reader with a little hands-on warm-up experience, a simple demo project has been provided. The code is exceptionally simple; it is comprised of no more than one header and two source files. However, it is carefully designed to illustrate the points of extraordinarily importance for understanding the broader picture.

The following files are the part of the project:

- Source file **main.c**, which contains the main() function.

- Header file **function.h**, which declares the functions called and the data accessed by the main() function.

- Source file **function.c**, which contains the source code implementations of functions and instantiation of the data referenced by the main() function.

The development environment used to build this simple project will be based on the **gcc** compiler running on Linux. Listings 2-1 through 2-3 contain the code used in the demo project.

Listing 2-1. function.h

```
#pragma once

#define FIRST_OPTION
#ifdef FIRST_OPTION
#define MULTIPLIER (3.0)
#else
#define MULTIPLIER (2.0)#endif

float add_and_multiply(float x, float y);
```

Listing 2-2. function.c

```
int nCompletionStatus = 0;

float add(float x, float y)
{
    float z = x + y;
    return z;
}

float add_and_multiply(float x, float y)
{
    float z = add(x,y);
    z *= MULTIPLIER;
    return z;
}
```

Listing 2-3. main.c

```
#include "function.h"
extern int nCompletionStatus = 0;
int main(int argc, char* argv[])
{
    float x = 1.0;
    float y = 5.0;
    float z;

    z = add_and_multiply(x,y);
    nCompletionStatus = 1;
    return 0;
}
```

Compiling

Once you have written your source code, it is the time to immerse yourself in the process of code building, whose mandatory first step is the compiling stage. Before going into the intricacies of compiling, a few simple introductory terms will be presented first.

Introductory Definitions

Compiling in the broad sense can be defined as the process of transforming source code written in one programming language into another programming language. The following set of introductory facts is important for your overall understanding of the compilation process:

- The process of compiling is performed by the program called the **compiler**.
- The input for the compiler is a **translation unit**. A typical translation unit is a text file containing the source code.
- A program is typically comprised of many translation units. Even though it is perfectly possible and legal to keep all the project's source code in a single file, there are good reasons (explained in the previous section) of why it is typically not the case.

11

- The output of the compilation is a collection of binary **object files**, one for each of the input translation units.

- In order to become suitable for execution, the object files need to be processed through another stage of program building called **linking**.

Figure 2-1 illustrates the concept of compiling.

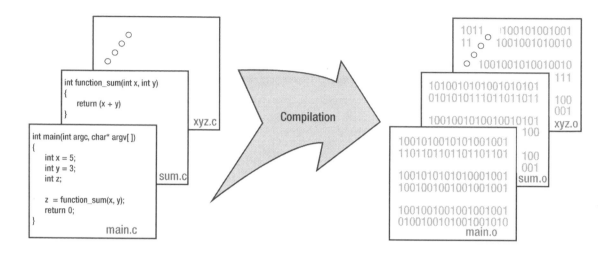

Figure 2-1. *The compiling stage*

Related Definitions

The following variety of compiler use cases is typically encountered:

- **Compilation** in the strict meaning denotes the process of translating the code of a higher-level language to the code of a lower-level language (typically, assembler or even machine code) production files.

- If the compilation is performed on one platform (CPU/OS) to produce code to be run on some other platform (CPU/OS), it is called **cross-compilation**. The usual practice is to use some of the desktop OSes (Linux, Windows) to generate the code for embedded or mobile devices.

- **Decompilation** (disassembling) is the process of converting the source code of a lower-level language to the higher-level language.

- **Language translation** is the process of transforming source code of one programming language to another programming language of the same level and complexity.

- **Language rewriting** is the process of rewriting the language expressions into a form more suitable for certain tasks (such as optimization).

The Stages of Compiling

The compilation process is not monolithic in nature. In fact, it can be roughly divided into the several stages (pre-processing, linguistic analysis, assembling, optimization, code emission), the details of which will be discussed next.

Preprocessing

The standard first step in processing the source files is running them through the special text processing program called a **preprocessor**, which performs one or more of the following actions:

- Includes the files containing definitions (include/header files) into the source files, as specified by the #include keyword.

- Converts the values specified by using #define statements into the constants.

- Converts the macro definitions into code at the variety of locations in which the macros are invoked.

- Conditionally includes or excludes certain parts of the code, based on the position of #if, #elif, and #endif directives.

The output of the preprocessor is the C/C++ code in its final shape, which will be passed to the next stage, syntax analysis.

Demo Project Preprocessing Example

The gcc compiler provides the mode in which only the preprocessing stage is performed on the input source files:

```
gcc -i <input file> -o <output preprocessed file>.i
```

Unless specified otherwise, the output of the preprocessor is the file that has the same name as the input file and whose file extension is **.i**. The result of running the preprocessor on the file function.c looks like that in Listing 2-4.

Listing 2-4. function.i

```
# 1 "function.c"
# 1 "
# 1 "
# 1 "function.h" 1

# 11 "function.h"
float add_and_multiply(float x, float y);
# 2 "function.c" 2

int nCompletionStatus = 0;

float add(float x, float y)
{
    float z = x + y;
    return z;
}

float add_and_multiply(float x, float y)
{
    float z = add(x,y);
    z *= MULTIPLIER;
    return z;
}
```

More compact and more meaningful preprocessor output may be obtained if few extra flags are passed to the gcc, like

```
gcc -E -P -i <input file> -o <output preprocessed file>.i
```

which results in the preprocessed file seen in Listing 2-5.

Listing 2-5. `function.i (Trimmed Down Version)`

```
float add_and_multiply(float x, float y);
int nCompletionStatus = 0;

float add(float x, float y)
{
    float z = x + y;
    return z;
}

float add_and_multiply(float x, float y)
{
    float z = add(x,y);
    z *= 3.0;
    return z;
}
```

Obviously, the preprocessor replaced the symbol MULTIPLIER, whose actual value, based on the fact that the USE_FIRST_OPTION variable was defined, ended up being 3.0.

Linguistic Analysis

During this stage, the compiler first converts the C/C++ code into a form more suitable for processing (eliminating comments and unnecessary white spaces, extracting tokens from the text, etc.). Such an optimized and compacted form of source code is lexically analyzed, with the intention of checking whether the program satisfies the syntax rules of the programming language in which it was written. If deviations from the syntax rules are detected, errors or warnings are reported. The errors are sufficient cause for the compilation to be terminated, whereas warnings may or may not be sufficient, depending on the user's settings.

More precise insight into this stage of the compilation process reveals three distinct stages:

- **Lexical analysis**, which breaks the source code into non-divisible tokens. The next stage,

- **Parsing/syntax analysis** concatenates the extracted tokens into the chains of tokens, and verifies that their ordering makes sense from the standpoint of programming language rules. Finally,

- **Semantic analysis** is run with the intent to discover whether the syntactically correct statements actually make any sense. For example, a statement that adds two integers and assigns the result to an object will pass syntax rules, but may not pass semantic check (unless the object has overridden assignment operator).

During the linguistic analysis stage, the compiler probably more deserves to be called "complainer," as it tends to more complain about typos or other errors it encounters than to actually compile the code.

Assembling

The compiler reaches this stage only after the source code is verified to contain no syntax errors. In this stage, the compiler tries to convert the standard language constructs into the constructs specific to the actual CPU instruction set. Different CPUs feature different functionality treats, and in general different sets of available instructions, registers, interrupts, which explains the wide variety of compilers for an even wider variety of processors.

Demo Project Assembling Example

The gcc compiler provides the mode of operation in which the input files' source code is converted into the ASCII text file containing the lines of assembler instructions specific to the chip and/or the operating system.

```
$ gcc -S <input file> -o <output assembler file>.s
```

Unless specified otherwise, the output of the preprocessor is the file that has the same name as the input file and whose file extension is **.s**.

The generated file is not suitable for execution; it is merely a text file carrying the human-readable mnemonics of assembler instructions, which can be used by the developer to get a better insight into the details of the inner workings of the compilation process.

In the particular case of the X86 processor architecture, the assembler code may conform to one of the two supported instruction printing formats,

- AT&T format
- Intel format

the choice of which may be specified by passing an extra command-line argument to the gcc assembler. The choice of format is mostly the matter of the developer's personal taste.

AT&T Assembly Format Example

When the file function.c is assembled into the AT&T format by running the following command

```
$ gcc -S -masm=att function.c -o function.s
```

it creates the output assembler file, which looks the code shown in Listing 2-6.

Listing 2-6. function.s (AT&T Assembler Format)

```
        .file       "function.c"
        .globl      nCompletionStatus
        .bss
        .align 4
        .type       nCompletionStatus, @object
        .size       nCompletionStatus, 4
nCompletionStatus:
        .zero       4
        .text
        .globl      add
        .type       add, @function
```

```
add:
.LFB0:
        .cfi_startproc
        pushl       %ebp
        .cfi_def_cfa_offset 8
        .cfi_offset 5, -8
        movl        %esp, %ebp
        .cfi_def_cfa_register 5
        subl        $20, %esp
        flds        8(%ebp)
        fadds       12(%ebp)
        fstps       -4(%ebp)
        movl        -4(%ebp), %eax
        movl        %eax, -20(%ebp)
        flds        -20(%ebp)
        leave
        .cfi_restore 5
        .cfi_def_cfa 4, 4
        ret
        .cfi_endproc
.LFE0:
        .size       add, .-add
        .globl      add_and_multiply
        .type       add_and_multiply, @function
add_and_multiply:
.LFB1:
        .cfi_startproc
        pushl       %ebp
        .cfi_def_cfa_offset 8
        .cfi_offset 5, -8
        movl        %esp, %ebp
        .cfi_def_cfa_register 5
        subl        $28, %esp
        movl        12(%ebp), %eax
        movl        %eax, 4(%esp)
        movl        8(%ebp), %eax
        movl        %eax, (%esp)
        call        add
        fstps       -4(%ebp)
        flds        -4(%ebp)
        flds        .LC1
        fmulp       %st, %st(1)
        fstps       -4(%ebp)
        movl        -4(%ebp), %eax
        movl        %eax, -20(%ebp)
        flds        -20(%ebp)
        leave
        .cfi_restore 5
        .cfi_def_cfa 4, 4
        ret
        .cfi_endproc
```

```
.LFE1:
      .size      add_and_multiply, .-add_and_multiply
      .section          .rodata
      .align 4
.LC1:
      .long      1077936128
      .ident     "GCC: (Ubuntu/Linaro 4.6.3-1ubuntu5) 4.6.3"
      .section          .note.GNU-stack,"",@progbits
```

Intel Assembly Format Example

The same file (function.c) may be assembled into the Intel assembler format by running the following command,

```
$ gcc -S -masm=intel function.c -o function.s
```

which results with the assembler file shown in Listing 2-7.

Listing 2-7. function.s (Intel Assembler Format)

```
      .file      "function.c"
      .intel_syntax noprefix
      .globl     nCompletionStatus
      .bss
      .align 4
      .type      nCompletionStatus, @object
      .size      nCompletionStatus, 4
nCompletionStatus:
      .zero      4
      .text
      .globl     add
      .type      add, @function
add:
.LFB0:
      .cfi_startproc
      push       ebp
      .cfi_def_cfa_offset 8
      .cfi_offset 5, -8
      mov        ebp, esp
      .cfi_def_cfa_register 5
      sub        esp, 20
      fld        DWORD PTR [ebp+8]
      fadd       DWORD PTR [ebp+12]
      fstp       DWORD PTR [ebp-4]
      mov        eax, DWORD PTR [ebp-4]
      mov        DWORD PTR [ebp-20], eax
      fld        DWORD PTR [ebp-20]
      leave
      .cfi_restore 5
      .cfi_def_cfa 4, 4
      ret
      .cfi_endproc
```

```
.LFE0:
        .size       add, .-add
        .globl      add_and_multiply
        .type       add_and_multiply, @function
add_and_multiply:
.LFB1:
        .cfi_startproc
        push        ebp
        .cfi_def_cfa_offset 8
        .cfi_offset 5, -8
        mov         ebp, esp
        .cfi_def_cfa_register 5
        sub         esp, 28
        mov         eax, DWORD PTR [ebp+12]
        mov         DWORD PTR [esp+4], eax
        mov         eax, DWORD PTR [ebp+8]
        mov         DWORD PTR [esp], eax
        call        add
        fstp        DWORD PTR [ebp-4]
        fld         DWORD PTR [ebp-4]
        fld         DWORD PTR .LC1
        fmulp       st(1), st
        fstp        DWORD PTR [ebp-4]
        mov         eax, DWORD PTR [ebp-4]
        mov         DWORD PTR [ebp-20], eax
        fld         DWORD PTR [ebp-20]
        leave
        .cfi_restore 5
        .cfi_def_cfa 4, 4
        ret
        .cfi_endproc
.LFE1:
        .size       add_and_multiply, .-add_and_multiply
        .section            .rodata
        .align 4
.LC1:
        .long       1077936128
        .ident      "GCC: (Ubuntu/Linaro 4.6.3-1ubuntu5) 4.6.3"
        .section            .note.GNU-stack,"",@progbits
```

Optimization

Once the first assembler version corresponding to the original source code is created, the optimization effort starts, in which usage of the registers is minimized. Additionally, the analysis may indicate that certain parts of the code do not in fact need to be executed, and such parts of the code are eliminated.

Code Emission

Finally, the moment has come to create the compilation output: **object files**, one for each translation unit. The assembly instructions (written in human-readable ASCII code) are at this stage converted into the binary values of the corresponding machine instructions (opcodes) and written to the specific locations in the object file(s).

The object file is still not ready to be served as the meal to the hungry processor. The reasons why are the essential topic of this whole book. The interesting topic at this moment is the analysis of an object file.

Being a binary file makes the object file substantially different than the outputs of preprocessing and assembling procedures, both of which are ASCII files, inherently readable by humans. The differences become the most obvious when we, the humans, try to take a closer look at the contents.

Other than obvious choice of using the hex editor (not very helpful unless you write compilers for living), a specific procedure called **disassembling** is taken in order to get a detailed insight into the contents of an object file.

On the overall path from the ASCII files toward the binary files suitable for execution on the concrete machine, the disassembling may be viewed as a little U-turn detour in which the almost-ready binary file is converted into the ASCII file to be served to the curious eyes of the software developer. Fortunately, this little detour serves only the purpose of supplying the developer with better orientation, and is normally not performed without a real cause.

Demo Project Compiling Example

The gcc compiler may be set to perform the complete compilation (preprocessing and assembling and compiling), a procedure that generates the binary object file (standard extension .o) whose structure follows the ELF format guidelines. In addition to usual overhead (header, tables, etc.), it contains all the pertinent sections (.text, .code, .bss, etc.). In order to specify the compilation only (no linking as of yet), the following command line may be used:

```
$ gcc -c <input file> -o <output file>.o
```

Unless specified otherwise, the output of the preprocessor is the file that has the same name as the input file and whose file extension is **.o**.

The content of the generated object file is not suitable for viewing in a text editor. The hex editor/viewer is a bit more suitable, as it will not be confused by the nonprintable characters and absences of newline characters. Figure 2-2 shows the binary contents of the object file function.o generated by compiling the file function.c of this demo project.

```
00000000  7f 45 4c 46 01 01 01 00   00 00 00 00 00 00 00 00   |.ELF............|
00000010  01 00 03 00 01 00 00 00   00 00 00 00 00 00 00 00   |................|
00000020  6c 01 00 00 00 00 00 00   34 00 00 00 00 00 28 00   |l.......4.....(.|
00000030  0d 00 0a 00 55 89 e5 83   ec 14 d9 45 08 d8 45 0c   |....U......E..E.|
00000040  d9 5d fc 8b 45 fc 89 45   ec d9 45 ec c9 c3 55 89   |.]..E..E..E...U.|
00000050  e5 83 ec 1c 8b 45 0c 89   44 24 04 8b 45 08 89 04   |.....E..D$..E...|
00000060  24 e8 fc ff ff ff d9 5d   fc d9 45 fc d9 05 00 00   |$......]..E.....|
00000070  00 00 de c9 d9 5d fc 8b   45 fc 89 45 ec d9 45 ec   |.....]..E..E..E.|
00000080  c9 c3 00 00 00 00 40 40   00 47 43 43 3a 20 28 55   |......@@.GCC: (U|
00000090  62 75 6e 74 75 2f 4c 69   6e 61 72 6f 20 34 2e 36   |buntu/Linaro 4.6|
000000a0  2e 33 2d 31 75 62 75 6e   74 75 35 29 20 34 2e 36   |.3-1ubuntu5) 4.6|
000000b0  2e 33 00 00 14 00 00 00   00 00 00 00 01 7a 52 00   |.3...........zR.|
000000c0  01 7c 08 01 1b 0c 04 04   88 01 00 00 1c 00 00 00   |.|..............|
000000d0  1c 00 00 00 00 00 00 00   1a 00 00 00 00 41 0e 08   |.............A..|
000000e0  85 02 42 0d 05 56 c5 0c   04 04 00 00 1c 00 00 00   |..B..V..........|
000000f0  3c 00 00 00 1a 00 00 00   34 00 00 00 00 41 0e 08   |<.......4....A..|
00000100  85 02 42 0d 05 70 c5 0c   04 04 00 00 00 2e 73 79   |..B..p........sy|
00000110  6d 74 61 62 00 2e 73 74   72 74 61 62 00 2e 73 68   |mtab..strtab..sh|
00000120  73 74 72 74 61 62 00 2e   72 65 6c 2e 74 65 78 74   |strtab..rel.text|
00000130  00 2e 64 61 74 61 00 2e   62 73 73 00 2e 72 6f 64   |..data..bss..rod|
00000140  61 74 61 00 2e 63 6f 6d   6d 65 6e 74 00 2e 6e 6f   |ata..comment..no|
00000150  74 65 2e 47 4e 55 2d 73   74 61 63 6b 00 2e 72 65   |te.GNU-stack..re|
00000160  6c 2e 65 68 5f 66 72 61   6d 65 00 00 00 00 00 00   |l.eh_frame......|
00000170  00 00 00 00 00 00 00 00   00 00 00 00 00 00 00 00   |................|
*
00000190  00 00 00 00 1f 00 00 00   01 00 00 00 06 00 00 00   |................|
000001a0  00 00 00 00 34 00 00 00   4e 00 00 00 00 00 00 00   |....4...N.......|
000001b0  00 00 00 00 04 00 00 00   00 00 00 00 1b 00 00 00   |................|
000001c0  09 00 00 00 00 00 00 00   00 00 00 00 68 04 00 00   |............h...|
000001d0  10 00 00 00 0b 00 00 00   01 00 00 00 04 00 00 00   |................|
000001e0  08 00 00 00 25 00 00 00   01 00 00 00 03 00 00 00   |....%...........|
000001f0  00 00 00 00 84 00 00 00   00 00 00 00 00 00 00 00   |................|
00000200  00 00 00 00 04 00 00 00   00 00 00 00 2b 00 00 00   |............+...|
00000210  08 00 00 00 03 00 00 00   00 00 00 00 84 00 00 00   |................|
00000220  04 00 00 00 00 00 00 00   00 00 00 00 04 00 00 00   |................|
00000230  00 00 00 00 30 00 00 00   01 00 00 00 02 00 00 00   |....0...........|
00000240  00 00 00 00 84 00 00 00   04 00 00 00 00 00 00 00   |................|
00000250  00 00 00 00 04 00 00 00   00 00 00 00 38 00 00 00   |............8...|
00000260  01 00 00 00 30 00 00 00   00 00 00 00 88 00 00 00   |....0...........|
00000270  2b 00 00 00 00 00 00 00   00 00 00 00 01 00 00 00   |+...............|
00000280  01 00 00 00 41 00 00 00   01 00 00 00 00 00 00 00   |....A...........|
00000290  00 00 00 00 b3 00 00 00   00 00 00 00 00 00 00 00   |................|
000002a0  00 00 00 00 01 00 00 00   00 00 00 00 55 00 00 00   |............U...|
000002b0  01 00 00 00 02 00 00 00   00 00 00 00 b4 00 00 00   |................|
000002c0  58 00 00 00 00 00 00 00   00 00 00 00 04 00 00 00   |X...............|
000002d0  00 00 00 00 51 00 00 00   09 00 00 00 00 00 00 00   |....Q...........|
000002e0  00 00 00 00 78 04 00 00   10 00 00 00 0b 00 00 00   |....x...........|
```

Figure 2-2. *Binary contents of an object file*

Obviously, merely taking a look at the hex values of the object file does not tell us a whole lot. The disassembling procedure has the potential to tell us far more.

The Linux tool called **objdump** (part of popular *binutils* package) specializes in disassembling the binary files, among a whole lot of other things. In addition to converting the sequence of binary machine instructions specific to a given platform, it also specifies the addresses at which the instructions reside.

It should not be a huge surprise that it supports both AT&T (default) as well as Intel flavors of printing the assembler code.

By running the simple form of objdump command,

```
$ objdump -D <input file>.o
```

you get the following contents printed on the terminal screen:

disassembled output of function.o (AT&T assembler format)
```
function.o:     file format elf32-i386

Disassembly of section .text:

00000000 <add>:
   0:   55                      push   %ebp
   1:   89 e5                   mov    %esp,%ebp
   3:   83 ec 14                sub    $0x14,%esp
   6:   d9 45 08                flds   0x8(%ebp)
   9:   d8 45 0c                fadds  0xc(%ebp)
   c:   d9 5d fc                fstps  -0x4(%ebp)
   f:   8b 45 fc                mov    -0x4(%ebp),%eax
  12:   89 45 ec                mov    %eax,-0x14(%ebp)
  15:   d9 45 ec                flds   -0x14(%ebp)
  18:   c9                      leave
  19:   c3                      ret

0000001a <add_and_multiply>:
  1a:   55                      push   %ebp
  1b:   89 e5                   mov    %esp,%ebp
  1d:   83 ec 1c                sub    $0x1c,%esp
  20:   8b 45 0c                mov    0xc(%ebp),%eax
  23:   89 44 24 04             mov    %eax,0x4(%esp)
  27:   8b 45 08                mov    0x8(%ebp),%eax
  2a:   89 04 24                mov    %eax,(%esp)
  2d:   e8 fc ff ff ff          call   2e <add_and_multiply+0x14>
  32:   d9 5d fc                fstps  -0x4(%ebp)
  35:   d9 45 fc                flds   -0x4(%ebp)
  38:   d9 05 00 00 00 00       flds   0x0
  3e:   de c9                   fmulp  %st,%st(1)
  40:   d9 5d fc                fstps  -0x4(%ebp)
  43:   8b 45 fc                mov    -0x4(%ebp),%eax
  46:   89 45 ec                mov    %eax,-0x14(%ebp)
  49:   d9 45 ec                flds   -0x14(%ebp)
  4c:   c9                      leave
  4d:   c3                      ret
```

```
Disassembly of section .bss:

00000000 <nCompletionStatus>:
   0:   00 00                   add     %al,(%eax)
        ...

Disassembly of section .rodata:

00000000 <.rodata>:
   0:   00 00                   add     %al,(%eax)
   2:   40                      inc     %eax
   3:   40                      inc     %eax

Disassembly of section .comment:

00000000 <.comment>:
   0:   00 47 43                add     %al,0x43(%edi)
   3:   43                      inc     %ebx
   4:   3a 20                   cmp     (%eax),%ah
   6:   28 55 62                sub     %dl,0x62(%ebp)
   9:   75 6e                   jne     79 <add_and_multiply+0x5f>
   b:   74 75                   je      82 <add_and_multiply+0x68>
   d:   2f                      das
   e:   4c                      dec     %esp
   f:   69 6e 61 72 6f 20 34    imul    $0x34206f72,0x61(%esi),%ebp
  16:   2e 36 2e 33 2d 31 75    cs ss xor %cs:%ss:0x75627531,%ebp
  1d:   62 75
  1f:   6e                      outsb   %ds:(%esi),(%dx)
  20:   74 75                   je      97 <add_and_multiply+0x7d>
  22:   35 29 20 34 2e          xor     $0x2e342029,%eax
  27:   36 2e 33 00             ss xor %cs:%ss:(%eax),%eax

Disassembly of section .eh_frame:

00000000 <.eh_frame>:
   0:   14 00                   adc     $0x0,%al
   2:   00 00                   add     %al,(%eax)
   4:   00 00                   add     %al,(%eax)
   6:   00 00                   add     %al,(%eax)
   8:   01 7a 52                add     %edi,0x52(%edx)
   b:   00 01                   add     %al,(%ecx)
   d:   7c 08                   jl      17 <.eh_frame+0x17>
   f:   01 1b                   add     %ebx,(%ebx)
  11:   0c 04                   or      $0x4,%al
  13:   04 88                   add     $0x88,%al
  15:   01 00                   add     %eax,(%eax)
  17:   00 1c 00                add     %bl,(%eax,%eax,1)
  1a:   00 00                   add     %al,(%eax)
  1c:   1c 00                   sbb     $0x0,%al
  1e:   00 00                   add     %al,(%eax)
  20:   00 00                   add     %al,(%eax)
```

```
22:   00 00                  add    %al,(%eax)
24:   1a 00                  sbb    (%eax),%al
26:   00 00                  add    %al,(%eax)
28:   00 41 0e               add    %al,0xe(%ecx)
2b:   08 85 02 42 0d 05      or     %al,0x50d4202(%ebp)
31:   56                     push   %esi
32:   c5 0c 04               lds    (%esp,%eax,1),%ecx
35:   04 00                  add    $0x0,%al
37:   00 1c 00               add    %bl,(%eax,%eax,1)
3a:   00 00                  add    %al,(%eax)
3c:   3c 00                  cmp    $0x0,%al
3e:   00 00                  add    %al,(%eax)
40:   1a 00                  sbb    (%eax),%al
42:   00 00                  add    %al,(%eax)
44:   34 00                  xor    $0x0,%al
46:   00 00                  add    %al,(%eax)
48:   00 41 0e               add    %al,0xe(%ecx)
4b:   08 85 02 42 0d 05      or     %al,0x50d4202(%ebp)
51:   70 c5                  jo     18 <.eh_frame+0x18>
53:   0c 04                  or     $0x4,%al
55:   04 00                  add    $0x0,%al
      ...
```

Similarly, by specifying the Intel flavor,

```
$ objdump -D -M intel <input file>.o
```

you get the following contents printed on the terminal screen:

disassembled output of function.o (Intel assembler format)
```
function.o:      file format elf32-i386

Disassembly of section .text:

00000000 <add&gt:
   0:   55                   push   ebp
   1:   89 e5                mov    ebp,esp
   3:   83 ec 14             sub    esp,0x14
   6:   d9 45 08             fld    DWORD PTR [ebp+0x8]
   9:   d8 45 0c             fadd   DWORD PTR [ebp+0xc]
   c:   d9 5d fc             fstp   DWORD PTR [ebp-0x4]
   f:   8b 45 fc             mov    eax,DWORD PTR [ebp-0x4]
  12:   89 45 ec             mov    DWORD PTR [ebp-0x14],eax
  15:   d9 45 ec             fld    DWORD PTR [ebp-0x14]
  18:   c9                   leave
  19:   c3                   ret
```

```
0000001a <add_and_multiply>:
  1a:  55                        push    ebp
  1b:  89 e5                     mov     ebp,esp
  1d:  83 ec 1c                  sub     esp,0x1c
  20:  8b 45 0c                  mov     eax,DWORD PTR [ebp+0xc]
  23:  89 44 24 04               mov     DWORD PTR [esp+0x4],eax
  27:  8b 45 08                  mov     eax,DWORD PTR [ebp+0x8]
  2a:  89 04 24                  mov     DWORD PTR [esp],eax
  2d:  e8 fc ff ff ff            call    2e <add_and_multiply+0x14>
  32:  d9 5d fc                  fstp    DWORD PTR [ebp-0x4]
  35:  d9 45 fc                  fld     DWORD PTR [ebp-0x4]
  38:  d9 05 00 00 00 00         fld     DWORD PTR ds:0x0
  3e:  de c9                     fmulp   st(1),st
  40:  d9 5d fc                  fstp    DWORD PTR [ebp-0x4]
  43:  8b 45 fc                  mov     eax,DWORD PTR [ebp-0x4]
  46:  89 45 ec                  mov     DWORD PTR [ebp-0x14],eax
  49:  d9 45 ec                  fld     DWORD PTR [ebp-0x14]
  4c:  c9                        leave
  4d:  c3                        ret
```

Disassembly of section .bss:

```
00000000 <nCompletionStatus>:
   0:  00 00                     add     BYTE PTR [eax],al
        ...
```

Disassembly of section .rodata:

```
00000000 <.rodata>:
   0:  00 00                     add     BYTE PTR [eax],al
   2:  40                        inc     eax
   3:  40                        inc     eax
```

Disassembly of section .comment:

```
00000000 <.comment>:
   0:  00 47 43                  add     BYTE PTR [edi+0x43],al
   3:  43                        inc     ebx
   4:  3a 20                     cmp     ah,BYTE PTR [eax]
   6:  28 55 62                  sub     BYTE PTR [ebp+0x62],dl
   9:  75 6e                     jne     79 <add_and_multiply+0x5f>
   b:  74 75                     je      82 <add_and_multiply+0x68>
   d:  2f                        das
   e:  4c                        dec     esp
   f:  69 6e 61 72 6f 20 34      imul    ebp,DWORD PTR [esi+0x61],0x34206f72
  16:  2e 36 2e 33 2d 31 75      cs ss xor ebp,DWORD PTR cs:ss:0x75627531
  1d:  62 75
  1f:  6e                        outs    dx,BYTE PTR ds:[esi]
  20:  74 75                     je      97 <add_and_multiply+0x7d>
  22:  35 29 20 34 2e            xor     eax,0x2e342029
  27:  36 2e 33 00               ss xor eax,DWORD PTR cs:ss:[eax]
```

```
Disassembly of section .eh_frame:

00000000 <.eh_frame>:
   0:	14 00                 adc    al,0x0
   2:	00 00                 add    BYTE PTR [eax],al
   4:	00 00                 add    BYTE PTR [eax],al
   6:	00 00                 add    BYTE PTR [eax],al
   8:	01 7a 52              add    DWORD PTR [edx+0x52],edi
   b:	00 01                 add    BYTE PTR [ecx],al
   d:	7c 08                 jl     17 < .eh_frame+0x17>
   f:	01 1b                 add    DWORD PTR [ebx],ebx
  11:	0c 04                 or     al,0x4
  13:	04 88                 add    al,0x88
  15:	01 00                 add    DWORD PTR [eax],eax
  17:	00 1c 00              add    BYTE PTR [eax+eax*1],bl
  1a:	00 00                 add    BYTE PTR [eax],al
  1c:	1c 00                 sbb    al,0x0
  1e:	00 00                 add    BYTE PTR [eax],al
  20:	00 00                 add    BYTE PTR [eax],al
  22:	00 00                 add    BYTE PTR [eax],al
  24:	1a 00                 sbb    al,BYTE PTR [eax]
  26:	00 00                 add    BYTE PTR [eax],al
  28:	00 41 0e              add    BYTE PTR [ecx+0xe],al
  2b:	08 85 02 42 0d 05     or     BYTE PTR [ebp+0x50d4202],al
  31:	56                    push   esi
  32:	c5 0c 04              lds    ecx,FWORD PTR [esp+eax*1]
  35:	04 00                 add    al,0x0
  37:	00 1c 00              add    BYTE PTR [eax+eax*1],bl
  3a:	00 00                 add    BYTE PTR [eax],al
  3c:	3c 00                 cmp    al,0x0
  3e:	00 00                 add    BYTE PTR [eax],al
  40:	1a 00                 sbb    al,BYTE PTR [eax]
  42:	00 00                 add    BYTE PTR [eax],al
  44:	34 00                 xor    al,0x0
  46:	00 00                 add    BYTE PTR [eax],al
  48:	00 41 0e              add    BYTE PTR [ecx+0xe],al
  4b:	08 85 02 42 0d 05     or     BYTE PTR [ebp+0x50d4202],al
  51:	70 c5                 jo     18 <.eh_frame+0x18>
  53:	0c 04                 or     al,0x4
  55:	04 00                 add    al,0x0
        ...
```

Object File Properties

The output of the compilation process is one or more binary object files, whose structure is the natural next topic of interest. As you will see shortly, the structure of object files contains many details of importance on the path of truly understanding the broader picture.

In a rough sketch,

- An object file is the result of translating its original corresponding source file. The result of compilation is the collection of as many object files as there are source files in the project.

    After the compiling completes, the object file keeps representing its original source file in subsequent stages of the program building process.

- The basic ingredients of an object file are the **symbols** (references to the memory addresses in program or data memory) as well as the **sections**.

    Among the sections most frequently found in the object files are the code (.text), initialized data (.data), uninitialized data (.bss), and some of the more specialized sections (debugging information, etc.).

- The ultimate intention behind the idea of building the program is that the sections obtained by compiling individual source files be combined (tiled) together into the single binary executable file.

    Such binary file would contain the sections of the same type (.text, .data, .bss, …) obtained by tiling together the sections from the individual files. Figuratively speaking, an object file can be viewed as a simple tile waiting to find its place in the giant mosaic of the process memory map.

- The inner structure of the object file does not, however, suggest where the individual sections will ultimately reside in the program memory map. For that reason, the address ranges of each section in each of the object files is tentatively set to start from a zero value.

    The actual address range at which a section from an object file will ultimately reside in the program map will be determined in the subsequent stages (linking) of program building process.

- In the process of tiling object files' sections into the resultant program memory map, the only truly important parameter is the length of its sections, or to say it more precisely, its address range.

- The object file carries no sections that would contribute to the stack and/or heap. The contents of these two sections of the memory map are completely determined at runtime, and other than the default byte length, require no program-specific initial settings.

- The object file's contribution to the program's .bss (uninitialized data) section is very rudimentary; the .bss section is described merely by its byte length. This meager information is just what is needed for the loader to establish the .bss section as a part of the memory in which some data will be stored.

In general, the information is stored in the object files according to a certain set of rules epitomized in the form of binary format specification, whose details vary across the different platforms (Windows vs. Linux, 32-bit vs. 64-bit, x86 vs. ARM processor family).

Typically, the binary format specifications are designed to support the C/C++ language constructs and the associated implementation problems. Very frequently, the binary format specification covers a variety of binary file modes such as executables, static libraries, and dynamic libraries.

On Linux, the Executable and Linkable Format (ELF) has gained the prevalence. On Windows, the binaries typically conform to the PE/COFF format specification.

Compilation Process Limitations

Step by step, the pieces of the gigantic puzzle of program building process are starting to fall in place, and the broad and clear picture of the whole story slowly emerges. So far, you've learned that the compilation process translates the ASCII source files into the corresponding collection of binary object files. Each of the object files contains sections, the destiny of each is to ultimately become a part of gigantic puzzle of the program's memory map, as illustrated in Figure 2-3.

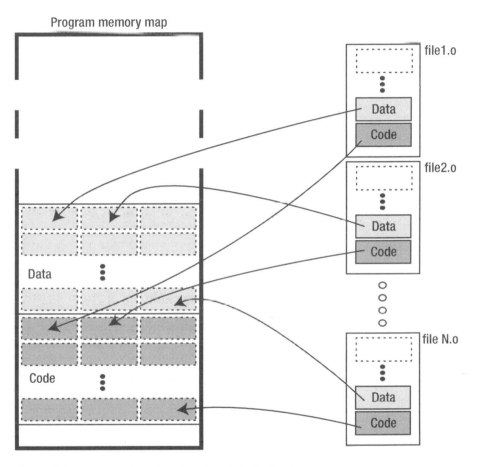

Figure 2-3. *Tiling the individual sections into the final program memory map*

The task that remains is to *tile the individual sections* stored across individual object files together into the body of program memory map. As mentioned in the previous sections, that task needs to be left to another stage of the program building process called **linking**.

The question that a careful observer can't help but asking (before going into the details of linking procedure) is *exactly why do we need a whole new stage of the building process*, or more precisely, *exactly why can't the compilation process described thus far complete the tiling part of the task?*

There are a few very solid reasons for splitting the build procedure, and the rest of this section will try to clarify the circumstances leading to such decision.

In short, the answer can be provided in a few simple statements. First, combining the sections together (especially the code sections) is not always simple. This factor definitely plays certain role, but is not sufficient; there are many programming languages whose program building process can be completed in one step (in other words, they do not require dividing the procedure into the two stages).

Second, the code reuse principle applied to the process of program building (and the ability to combine together the binary parts coming from various projects) definitely affirmed the decision to implement the C/C++ building as a two-step (compiling and linking) procedure.

What Makes Section Combining so Complicated?

For the most part, the translation of source code into binary object files is a fairly simple process. The lines of code are translated into processor-specific machine code instructions; the space for initialized variables is reserved and initial values are written to it; the space for uninitialized variables is reserved and filled out with zeros, etc.

However, there is a part of the whole story which is bound to cause some problems: even though the source code is grouped into the dedicated source files, being part of the same program implies that certain mutual connections must exist. Indeed, the connections between the distinct parts of the code are typically established through either the following two options:

- *Function calls between functionally separate bodies of code*:

  For example, a function in the GUI-related source file of a chat application may call a function in the TCP/IP networking source file, which in turn may call a function located in the encryption source file.

- *External variables*:

  In the domain of the C programming language (substantially less in the C++ domain), it was a usual practice to reserve globally visible variables to maintain the state of interest for various parts of code. A variable intended for broader use is typically declared in one source file as global variable, and referenced from all other source files as extern variable.

  A typical example is the errno variable used in standard C libraries to keep the value of the last encountered error.

In order to access either of the two (which are commonly referred to as **symbols**), their addresses (more precisely, the function's address in the program memory and/or the global variable's address in data memory) must be known.

However, the actual address cannot be known before the individual sections are incorporated into the corresponding program section (i.e., before the section tiling is completed!!!). Until then, a meaningful connection between a function and its caller and/or access to the external variable is impossible to establish, which are both suitably reported as unresolved references. Please notice that this problem does not happen when the function or global variable is referenced from the same source file in which it was defined. In this particular case, both the function/variable and their caller/user end up being the part of the same section, and their positions relative to each other are known before the "grand puzzle completion." In such cases, as soon as the tiling of the sections is completed, the relative memory addresses become concrete and usable.

As mentioned earlier in this section, solving this kind of problem still does not mandate that a build procedure must be divided into two distinct stages. As a matter of fact, many different languages successfully implement a one-pass build procedure. However, the concept of reusing (binary reusing in this case) applied to the realm of building the program (and the concept of libraries) ultimately confirms the decision to split the program building into the two stages (compiling and linking).

Linking

The second stage of the program building process is **linking**. The input to the linking process is the collection of object files created by the previously completed compiling stage. Each object file can be viewed as binary storage of individual source file contributions to the program memory map sections of all kinds (code, initialized data, uninitialized data, debugging information, etc.). The ultimate task of the linker is to form the resultant program memory map section out of individual contributions and to resolve all the references. As a reminder, the concept of virtual memory simplified the task of linker inasmuch as allowing it to assume that the program memory map that the linker needs to populate is a zero-based address range of identical size for each and every program, regardless of what address range the process will be given by the operating system at runtime.

For the sake of simplicity, I will cover in this example the simplest possible case, in which the contributions to the program memory map sections come solely from the files belonging to the same project. In reality, due to advancement of binary reuse concept, this may not be true.

Linking Stages

The linking process happens through a sequence of stages (relocation, reference resolving), which will be discussed in detail next.

Relocation

The first stage of a linking procedure is nothing else than tiling, a process in which sections of various kinds contained in individual object files are combined together to create the program memory map sections (see Figure 2-4). In order to complete this task, the previously neutral, zero-based address ranges of contributing sections get translated into the more concrete address ranges of resultant program memory map.

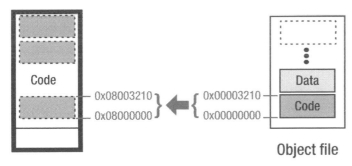

Program memory map

Figure 2-4. *Relocation, the first phase of the linking stage*

The wording "more concrete" is used to emphasize the fact that the resultant program image created by the linker is still neutral by itself. Remember, the mechanism of virtual addressing makes it possible that each and every program has the same, identical, simple view of the program address space (which resides between 0 and 2^N), whereas the real physical address at which the program executes gets determined at runtime by the operating system, invisible to the program and programmer.

Once the relocation stage completes, most (but not all!) of the program memory map has been created.

Resolving References

Now comes the hard part. Taking sections, linearly translating their address ranges into the program memory map address ranges was fairly easy task. A much harder task is to establish the required connections between the various parts of the code, thus making the program homogenous.

Let's assume (rightfully so, given the simplicity of this demo program) that all the previous build stages (complete compilation as well as section relocation) have completed successfully. Now is the moment to point out exactly which kinds of problems are left for the last linking stage to resolve.

As mentioned earlier, the root cause of linking problems is fairly simple: pieces of code originated from different translation units (i.e., source files) and are trying to reference each other, but cannot possibly know where in memory these items will reside up until the object files are tiled into the body of program memory map. The components of the code that cause the most problems are the ones tightly bound to the address in either program memory (function entry points) or in data memory (global/static/extern) variables.

In this particular code example, you have the following situation:

- The function add_and_multiply calls the function add, which resides in the same source file (i.e., the same translation unit in the same object file). In this case, the address in the program memory of function add() is to some extent a known quantity and can be expressed by its relative offset of the code section of the object file function.o.

- Now function main calls function add_and_multiply and also references the extern variable nCompletionStatus and has huge problems figuring out the actual program memory address at which they reside. In fact, it only may assume that both of these *symbols* will at some point in the future reside *somewhere* in the process memory map. But, until the memory map is formed, two items cannot be considered as nothing else than *unresolved references*.

The situation is graphically described in Figure 2-5.

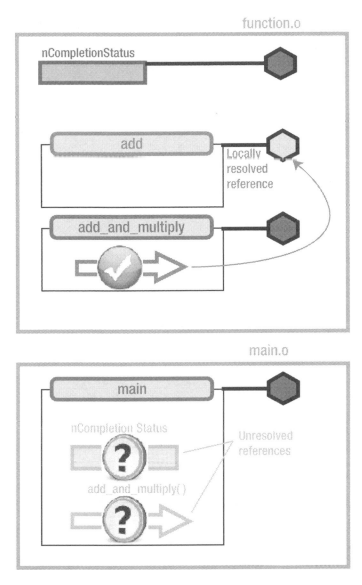

Figure 2-5. *The problem of unresolved references in its essential form*

In order to solve these kinds of problems, a linking stage of resolving the references must happen. What linker needs to do in this situation is to

- Examine the sections already tiled together in the program memory map.

- Find out which part of the code makes calls outside of its original section.

- Figure out where exactly (at which address in the memory map) the referenced part of the code resides.

- And finally, resolve the references by replacing dummy addresses in the machine instructions with the actual addresses of the program memory map.

Once the linker completes its magic, the situation may look like Figure 2-6.

Program memory map

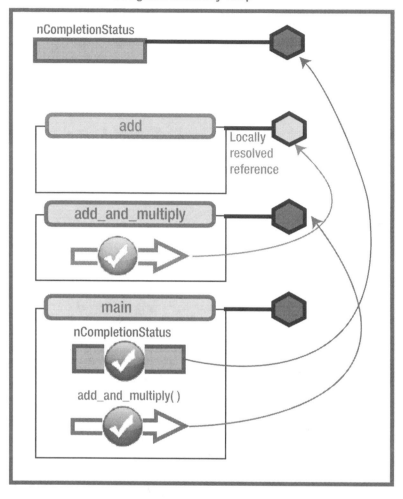

Figure 2-6. Resolved references

Demo Project Linking Example

There are two ways to compile and link the complete demo project to create the executable file so that it's ready for running.

In the step-by-step approach, you will first invoke the compiler on both of the source files to produce the object files. In the subsequent step, you will link both object files into the output executable.

```
$ gcc -c function.c main.c
$ gcc function.o main.o -o demoApp
```

In the all-at-once approach, the same operation may be completed by invoking the compiler and linker with just one command.

```
$ gcc function.c main.c -o demoApp
```

For the purposes of this demo, let's take the step-by-step approach, as it will generate the main.o object file, which contains very important details that I want to demonstrate here.

The disassembling of the file main.o,

```
$ objdump -D -M intel main.o
```

reveals that it contains unresolved references.

disassembled output of main.o (Intel assembler format)
```
main.o:     file format elf32-i386

Disassembly of section .text:

00000000 <main>:
   0:   55                      push   ebp
   1:   89 e5                   mov    ebp,esp
   3:   83 e4 f0                and    esp,0xfffffff0
   6:   83 ec 20                sub    esp,0x20
   9:   b8 00 00 80 3f          mov    eax,0x3f800000
   e:   89 44 24 14             mov    DWORD PTR [esp+0x14],eax
  12:   b8 00 00 a0 40          mov    eax,0x40a00000
  17:   89 44 24 18             mov    DWORD PTR [esp+0x18],eax
  1b:   8b 44 24 18             mov    eax,DWORD PTR [esp+0x18]
  1f:   89 44 24 04             mov    DWORD PTR [esp+0x4],eax
  23:   8b 44 24 14             mov    eax,DWORD PTR [esp+0x14]
  27:   89 04 24                mov    DWORD PTR [esp],eax
  2a:   e8 fc ff ff ff          call   2b <main + 0x2b>
  2f:   d9 5c 24 1c             fstp   DWORD PTR [esp+0x1c]
  33:   c7 05 00 00 00 00 01    mov    DWORD PTR ds:0x0,0x1
  3a:   00 00 00
  3d:   b8 00 00 00 00          mov    eax,0x0
  42:   c9                      leave
  43:   c3                      ret    :
```

Line 2a features a call instruction that jumps to itself (strange, isn't it?) whereas line 33 features the access of the variable residing at the address 0x0 (even more strange). Obviously, these two obviously strange values were inserted by the linker purposefully.

The disassembled output of the output executable, however, shows that not only the contents of the main.o object file have been relocated to the address range starting at the address 0x08048404, but also these two troubled spots have been resolved by the linker.

```
$ objdump -D -M intel demoApp
```

disassembled output of demoApp (Intel assembler format)
```
080483ce <add_and_multiply>:
 80483ce:       55                      push    ebp
 80483cf:       89 e5                   mov     ebp,esp
 80483d1:       83 ec 1c                sub     esp,0x1c
 80483d4:       8b 45 0c                mov     eax,DWORD PTR [ebp+0xc]
 80483d7:       89 44 24 04             mov     DWORD PTR [esp+0x4],eax
 80483db:       8b 45 08                mov     eax,DWORD PTR [ebp+0x8]
 80483de:       89 04 24                mov     DWORD PTR [esp],eax
 80483e1:       e8 ce ff ff ff          call    80483b4 <add>
 80483e6:       d9 5d fc                fstp    DWORD PTR [ebp-0x4]
 80483e9:       d9 45 fc                fld     DWORD PTR [ebp-0x4]
 80483ec:       d9 05 20 85 04 08       fld     DWORD PTR ds:0x8048520
 80483f2:       de c9                   fmulp   st(1),st
 80483f4:       d9 5d fc                fstp    DWORD PTR [ebp-0x4]
 80483f7:       8b 45 fc                mov     eax,DWORD PTR [ebp-0x4]
 80483fa:       89 45 ec                mov     DWORD PTR [ebp-0x14],eax
 80483fd:       d9 45 ec                fld     DWORD PTR [ebp-0x14]
 8048400:       c9                      leave
 8048401:       c3                      ret
 8048402:       90                      nop
 8048403:       90                      nop

08048404 <main>:
 8048404:       55                      push    ebp
 8048405:       89 e5                   mov     ebp,esp
 8048407:       83 e4 f0                and     esp,0xfffffff0
 804840a:       83 ec 20                sub     esp,0x20
 804840d:       b8 00 00 80 3f          mov     eax,0x3f800000
 8048412:       89 44 24 14             mov     DWORD PTR [esp+0x14],eax
 8048416:       b8 00 00 a0 40          mov     eax,0x40a00000
 804841b:       89 44 24 18             mov     DWORD PTR [esp+0x18],eax
 804841f:       8b 44 24 18             mov     eax,DWORD PTR [esp+0x18]
 8048423:       89 44 24 04             mov     DWORD PTR [esp+0x4],eax
 8048427:       8b 44 24 14             mov     eax,DWORD PTR [esp+0x14]
 804842b:       89 04 24                mov     DWORD PTR [esp],eax
 804842e:       e8 9b ff ff ff          call    80483ce <add_and_multiply>
 8048433:       d9 5c 24 1c             fstp    DWORD PTR [esp+0x1c]
 8048437:       c7 05 18 a0 04 08 01    mov     DWORD PTR ds:0x804a018,0x1
 804843e:       00 00 00
 8048441:       b8 00 00 00 00          mov     eax,0x0
 8048446:       c9                      leave   t:
```

The line at the memory map address 0x8048437 references the variable residing at address 0x804a018. The only open question now is what resides at that particular address?

The versatile objdump tool may help you get the answer to that question (a decent part of subsequent chapters is dedicated to this exceptionally useful tool).

By running the following command

```
$ objdump -x -j .bss demoApp
```

you can disassemble the .bss section carrying the uninitialized data, which reveals that your variable nCompletionStatus resides exactly at the address 0x804a018, as shown in Figure 2-7.

```
milan@milan$ objdump -x -j .bss demoApp
                                        o
                                        o
SYMBOL TABLE:                           o
0804a010 l    d  .bss    00000000              .bss
0804a010 l    O  .bss    00000001              completed.6159
0804a014 l    O  .bss    00000004              dtor_idx.6161
0804a018 g    O  .bss    00000004              nCompletionStatus
```

Figure 2-7. bss disassembled

Linker's Viewpoint

"When you've got a hammer in your hand, everything looks like a nail"

—Handy Hammer Syndrome

But seriously, folks....

Now that you know the intricacies of the linking task, it helps to zoom out a little bit and try to summarize the philosophy that guides the linker while running its usual tasks. As a matter of fact, the linker is a specific tool, which, unlike its older brother the compiler, is not interested in the minute details of the written code. Instead, it views the world as a set of object files that (much like puzzle pieces) are about to be combined together in a wider picture of program memory map, as illustrated by Figure 2-8.

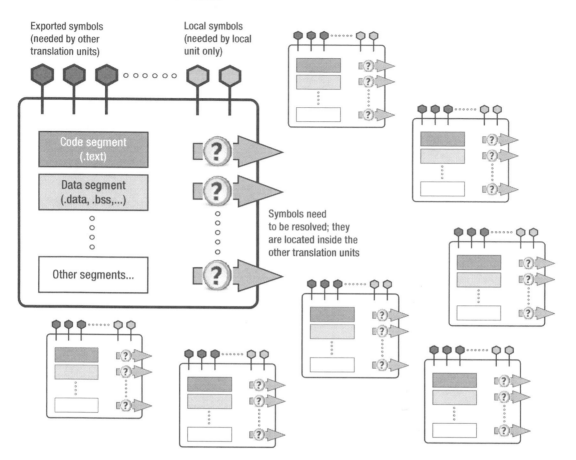

Exported symbols
(needed by other
translation units)

Local symbols
(needed by local
unit only)

Code segment
(.text)

Data segment
(.data, .bss,...)

Other segments...

Symbols need
to be resolved; they
are located inside the
other translation units

Figure 2-8. *The linker's view of the world*

It does not take a whole lot of imagination to find out that Figure 2-8 has a lot of resemblance with the left part of Figure 2-9, whereas the linker's ultimate task could be represented by the right part of the same figure.

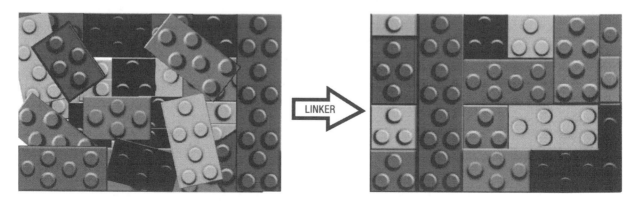

LINKER

Figure 2-9. *Linker's view of the world as seen by humans*

Executable File Properties

The ultimate result of the linking procedure is the binary executable file, whose layout follows the rules of the executable format suitable for the target platform. Regardless of the actual format differences, the executable file typically contains the resultant sections (.text, .data, .bss, and many more narrowly specialized ones) created by combining the contributions from individual object files. Most notably, the code (.text) section not only contains the individual tiles from object files, but the linker modified it to make sure that all the references between the individual tiles have been resolved, so that the function calls between different parts of code as well as variable accesses are accurate and meaningful.

Among all the symbols contained in the executable file, a very unique place belongs to the main function, as from the standpoint of C/C++ programs it is the function from which the entire program execution starts. However, this is not the very first part of code that executes when the program starts.

An exceptionally important detail that needs to be pointed out is that the executable file is not entirely made of code compiled from the project source files. As a matter of fact, a strategically important piece of code responsible for starting the program execution is added at the linking stage to the program memory map. This object code, which linker typically stores at the beginning of the program memory map, comes in two variants:

- **crt0** is the "plain vanilla" entry point, the first part of program code that gets executed under the control of kernel.

- **crt1** is the more modern startup routine with support for tasks to be completed before the main function gets executed and after the program terminates.

Having these details in mind, the overall structure of the program executable may be symbolically represented by Figure 2-10.

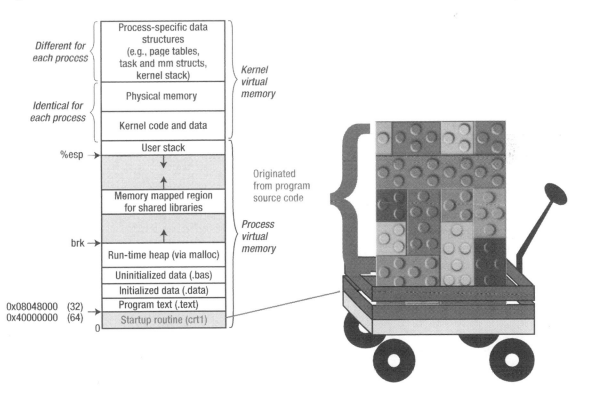

Figure 2-10. *Overall structure of an executable file*

As you will see later in the chapter devoted to dynamic libraries and dynamic linking, this extra piece of code, which gets provided by the operating system, makes almost all the difference between an executable and a dynamic library; the latter does not have this particular part of the code.

More details of the sequence of steps happening when a program execution starts will be discussed in the next chapter.

Variety of Section Types

Much like running an automobile cannot be imagined without the motor and a set of four wheels, executing the program cannot be imagined without the code (.text) and the data (.data and/or .bss) sections. These ingredients are naturally the quintessential part of the most basic program functionality.

However, much like the automobile is not only the motor and four wheels, the binary file contains many more sections. In order to finely synchronize the variety of operational tasks, the linker creates and inserts into the binary file many more different section types.

By convention, the section name starts with the dot (.) character. The names of the most important section types are platform independent; they are called the same regardless of the platform and the binary format it belongs to.

Throughout the course of this book, the meanings and roles of certain section types in the overall scheme of things will be discussed at length. Hopefully, by the time the book is read through, the reader will have a substantially wider and more focused understanding of the binary file sections.

In Table 2-1, the specification of the Linux' prevalent ELF binary format brings the following (http://man7.org/linux/man-pages/man5/elf.5.html) list of various section types provided in the alphabetical order. Even though the descriptions of individual sections are a bit meager, taking a glance at the variety of sections at this point may give reader fairly good idea about the variety available.

Table 2-1. *Linker Section Types*

Section Name	Description
.bss	This section holds the uninitialized data that contributes to the program's memory image. By definition, the system initializes the data with zeros when the program begins to run. This section is of type SHT_NOBITS. The attribute types are SHF_ALLOC and SHF_WRITE.
.comment	This section holds version control information. This section is of type SHT_PROGBITS. No attribute types are used.
.ctors	This section holds initialized pointers to the C++ constructor functions. This section is of type SHT_PROGBITS. The attribute types are SHF_ALLOC and SHF_WRITE.
.data	This section holds initialized data that contributes to the program's memory image. This section is of type SHT_PROGBITS. The attribute types are SHF_ALLOC and SHF_WRITE.
.data1	This section holds initialized data that contributes to the program's memory image. This section is of type SHT_PROGBITS. The attribute types are SHF_ALLOC and SHF_WRITE.
.debug	This section holds information for symbolic debugging. Thecontents are unspecified. This section is of type SHT_PROGBITS. No attribute types are used.
.dtors	This section holds initialized pointers to the C++ destructor functions. This section is of type SHT_PROGBITS. The attribute types are SHF_ALLOC and SHF_WRITE.
.dynamic	This section holds dynamic linking information. The section's attributes include the SHF_ALLOC bit. Whether the SHF_WRITE bit is set is processor-specific. This section is of type SHT_DYNAMIC. See the attributes above

(continued)

Table 2-1. (*continued*)

Section Name	Description
.dynstr	This section holds the strings needed for dynamic linking, most commonly the strings that represent the names associated with symbol table entries. This section is of type SHT_STRTAB. The attribute type used is SHF_ALLOC.
.dynsym	This section holds the dynamic linking symbol table. This section is of type SHT_DYNSYM. The attribute used is SHF_ALLOC.
.fini	This section holds executable instructions that contribute to the process termination code. When a program exits normally, the system arranges to execute the code in this section. This section is of type SHT_PROGBITS. The attributes used are SHF_ALLOC and SHF_EXECINSTR.
.gnu.version	This section holds the version symbol table, an array of ElfN_Half elements. This section is of type SHT_GNU_versym. The attribute type used is SHF_ALLOC.
.gnu.version_d	This section holds the version symbol definitions, a table of ElfN_Verdef structures. This section is of type SHT_GNU_verdef. The attribute type used is SHF_ALLOC.
.gnu_version.r	This section holds the version symbol needed elements, a table of ElfN_Verneed structures. This section is of type SHT_GNU_versym. The attribute type used is SHF_ALLOC.
.got	This section holds the global offset table. This section is of type SHT_PROGBITS. The attributes are processor-specific.
.got.plt	This section holds the procedure linkage table. This section is of type SHT_PROGBITS. The attributes are processor-specific.
.hash	This section holds a symbol hash table. This section is of type SHT_HASH. The attribute used is SHF_ALLOC.
.init	This section holds executable instructions that contribute to the process initialization code. When a program starts to run, the system arranges to execute the code in this section before calling the main program entry point. This section is of type SHT_PROGBITS. The attributes used are SHF_ALLOC and SHF_EXECINSTR.
.interp	This section holds the pathname of a program interpreter. If the file has a loadable segment that includes the section, the section's attributes will include the SHF_ALLOC bit. Otherwise, that bit will be off. This section is of type SHT_PROGBITS.
.line	This section holds the line number information for symbolic debugging, which describes the correspondence between the program source and the machine code. The contents are unspecified. This section is of type SHT_PROGBITS. No attribute types are used.
.note	This section holds information in the "Note Section" format. This section is of type SHT_NOTE. No attribute types are used. OpenBSD native executables usually contain a .note.openbsd.ident section to identify themselves, for the kernel to bypass any compatibility ELF binary emulation tests when loading the file.
.note.GNU-stack	This section is used in Linux object files for declaring stack attributes. This section is of type SHT_PROGBITS. The only attribute used is SHF_EXECINSTR. This indicates to the GNU linker that the object file requires an executable stack.
.plt	This section holds the procedure linkage table. This section is of type SHT_PROGBITS. The attributes are processor-specific.

(*continued*)

Table 2-1. (*continued*)

Section Name	Description
.relNAME	This section holds relocation information as described below. If the file has a loadable segment that includes relocation, the section's attributes will include the SHF_ALLOC bit. Otherwise, the bit will be off. By convention, "NAME" is supplied by the section to which the relocations apply. Thus a relocation section for .text normally would have the name .rel.text. This section is of type SHT_REL.
.relaNAME	This section holds relocation information as described below. If the file has a loadable segment that includes relocation, the section's attributes will include the SHF_ALLOC bit. Otherwise, the bit will be off. By convention, "NAME" is supplied by the section to which the relocations apply. Thus, a relocation section for .text normally would have the name .rela.text. This section is of type SHT_RELA.
.rodata	This section holds read-only data that typically contributes to a nonwritable segment in the process image. This section is of type SHT_PROGBITS. The attribute used is SHF_ALLOC.
.rodata1	This section holds read-only data that typically contributes to a nonwritable segment in the process image. This section is of type SHT_PROGBITS. The attribute used is SHF_ALLOC.
.shrstrtab	This section holds section names. This section is of type SHT_STRTAB. No attribute types are used.
.strtab	This section holds strings, most commonly the strings that represent the names associated with symbol table entries. If the file has a loadable segment that includes the symbol string table, the section's attributes will include the SHF_ALLOC bit. Otherwise the bit will be off. This section is of type SHT_STRTAB.
.symtab	This section holds a symbol table. If the file has a loadable segment that includes the symbol table, the section's attributes will include the SHF_ALLOC bit. Otherwise, the bit will be off. This section is of type SHT_SYMTAB.
.text	This section holds the "text," or executable instructions, of a program. This section is of type SHT_PROGBITS. The attributes used are SHF_ALLOC and SHF_EXECINSTR.

A Variety of Symbol Types

The ELF format provides a vast variety of linker symbol types, far larger than you can imagine at this early stage on your path toward understanding the intricacies of the linking process. At the present moment, you can clearly distinguish that symbols can be of either local scope or of broader visibility, typically needed by the other modules. Throughout the book material later on, the various symbol types will be discussed in substantially more detail.

Table 2-2 features the variety of symbol types, as shown in the man pages (http://linux.die.net/man/1/nm) of the useful ***nm*** symbol examination utility program. As a general rule, unless explicitly indicated (like in the case of "U" vs. "u"), the small letter denotes local symbols, whereas the capital letter indicates better symbol visibility (extern, global).

Table 2-2. *Linker Symbol Types*

Symbol Type	Description
"A"	The symbol's value is absolute, and will not be changed by further linking.
"B" or "b"	The symbol is in the uninitialized (.bss) data section.
"C"	The symbol is common. Common symbols are uninitialized data. When linking, multiple common symbols may appear with the same name. If the symbol is defined anywhere, the common symbols are treated as undefined references.
"D" or "d"	The symbol is in the initialized data section.
"G" or "g"	The symbol is in an initialized data section for small objects. Some object file formats permit more efficient access to small data objects, such as a global int variable as opposed to a large global array.
"i"	For PE format files, this indicates that the symbol is in a section specific to the implementation of DLLs. For ELF format files, this indicates that the symbol is an indirect function. This is a GNU extension to the standard set of ELF symbol types. It indicates a symbol that, if referenced by a relocation, does not evaluate to its address, but instead must be invoked at runtime. The runtime execution willthen return the value to be used in the relocation.
"N"	The symbol is a debugging symbol.
"p"	The symbols is in a stack unwind section.
"R" or "r"	The symbol is in a read only data section.
"S" or "s"	The symbol is in an uninitialized data section for small objects.
"T" or "t"	The symbol is in the text (code) section.
"U"	The symbol is undefined. In fact, this binary does not define this symbol, but expects that it eventually appears as the result of loading the dynamic libraries.
"u"	The symbol is a unique global symbol. This is a GNU extension to the standard set of ELF symbol bindings. For such a symbol, the dynamic linker will make sure that in the entire process there is just one symbol with this name and type in use.
"V" or "v"	The symbol is a weak object. When a weak defined symbol is linked with a normal defined symbol, the normal defined symbol is used with no error. When a weak undefined symbol is linked and the symbol is not defined, the value of the weak symbol becomes zero with no error. On some systems, uppercase indicates that a default value has been specified.
"W" or "w"	The symbol is a weak symbol that has not been specifically tagged as a weak object symbol. When a weak defined symbol is linked with a normal defined symbol, the normal definedsymbol is used with no error. When a weak undefined symbol is linked and the symbol is not defined, the value of the symbol is determined in a system-specific manner without error.On some systems, uppercase indicates that a default value has been specified.
"-"	The symbol is a stabs symbol in an a.out object file. In this case, the next values printed are the stabs other field, the stabs desc field, and the stab type. Stabs symbols are usedto hold debugging information.
"?"	The symbol type is unknown, or object file format-specific.

■ ■ ■

## CHAPTER 3

Program Execution Stages

The purpose of this chapter is to describe the sequence of events that happens when the user starts a program. The analysis is primarily focused on pointing out the details of interplay between the operating system and the layout of the executable binary file, which is tightly connected with the process memory map. Needless to say, the primary focus of this discussion is the execution sequence of the executable binaries created by building the code in C/C++.

Importance of the Shell

The program execution under the user's control typically happens through a shell, a program that monitors a user's actions on the keyboard and mouse. Linux features many different shells, the most popular being **sh, bash,** and **tcsh**.

Once the user types in the name of the command and presses the Enter key, the shell first tries to compare the typed command name against the one of its own built-in commands. If the program name is confirmed to not be any of the shell's supported commands, the shell tries to locate the binary whose name matches the command string. If user typed in only the program name (i.e. not the full path to the executable binary), the shell tries to locate the executable in each of the folders specified by the PATH environment variable. Once the full path of executable binary is known, the shell activates the procedure for loading and executing the binary.

The mandatory first action of the shell is to create a clone of itself by forking the identical child process. Creating the new process memory map by copying the shell's existing memory map seems like a strange move, as it is very likely that a new process memory map will have nothing in common with the memory map of the shell. This strange maneuver is made for a good reason: this way the shell effectively passes all of its environment variables to the new process. Indeed, soon after the new process memory map is created, the majority of its original contents get erased/zeroed out (except the part carrying the inherited environment variables) and overwritten with the memory map of the new process, which becomes ready for the execution stage. Figure 3-1 illustrates the idea.

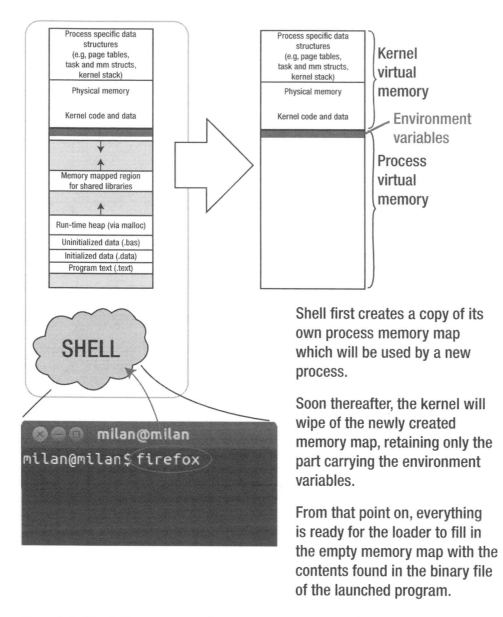

Shell first creates a copy of its own process memory map which will be used by a new process.

Soon thereafter, the kernel will wipe of the newly created memory map, retaining only the part carrying the environment variables.

From that point on, everything is ready for the loader to fill in the empty memory map with the contents found in the binary file of the launched program.

Figure 3-1. *The shell starts creating the new process memory map by copying its own process memory map, with the intention to pass its own environment variables to the new process*

From this point on, the shell may follow one of two possible scenarios. By default, the shell waits for its forked clone process to complete the command (i.e. that the launched program completes the execution). Alternatively, if the user types in an ampersand after the program name, the child process will be pushed to the background, and the shell will continue monitoring the user's subsequently typed commands. The very same mode may be achieved by the user not appending the ampersand after the executable name; instead, after the program is started, the user may press Ctrl-Z (which issues the SIGSTOP signal to the child process) and immediately after type "bg" (which issues the SIGCONT signal to the child process) in the shell window, which will cause the identical effect (pushing the shell child process to the background).

A very similar scenario of starting the program happens when user applies a mouse click to the application icon. The program that provides the icon (like a gnome-session and/or Nautilus File Explorer on Linux) takes the responsibility of translating the mouse click into the system( ) call, which causes a very similar sequence of events to happen as if the app were invoked by typing into the shell window.

Kernel Role

As soon as the shell delegates the task of running the program, the kernel reacts by invoking a function from the exec family of functions, all of which provide pretty much the same functionality, but differ in the details of how the execution parameters are specified. Regardless of which particular exec-type function is chosen, each of them ultimately makes a call to the sys_execve function, which starts the actual job of executing the program.

The immediate next step (which happens in function search_binary_handler (file fs/exec.c) is to identify the executable format. In addition to supporting the most recent ELF binary executable format, Linux provides backwards compatibility by supporting several other binary formats. If the ELF format is identified, the focus of action moves into the load_elf_binary function (file fs/binfmt_elf.c).

After the executable format is identified as one of the supported formats, the effort of preparing the process memory map for the execution commences. In particular, the child process created by the shell (a clone of the shell itself) is passed from shell to the kernel with the following intentions:

- The kernel obtains the sandbox (the process environment) and, more importantly, the associated memory, which can be used to launch the new program.

  The first thing that kernel will do is to completely wipe off most of the memory map. Immediately after, it will delegate to the loader the process of populating the wiped off memory map with the data read from the new program's binary executable fileSharePoint.

- By cloning the shell process (through the fork( ) call), the environment variables defined in the shell are passed onto the child process, which helps that the chain of environment variables' inheritance not get broken.

Loader Role

Before going into the details of loader functionality, it is important to point out that the loader and linker have somewhat different perspectives on the contents of the binary file.

Loader-Specific View of a Binary File (Sections vs. Segments)

The linker can be thought of as a highly sophisticated module capable of precisely distinguishing a wide variety of sections of various natures (code, uninitialized data, initialized data, constructors, debugging information, etc.). In order to resolve the references, it must intimately know the details of their internal structure.

On the other hand, the loader's responsibilities are far simpler. For the most part, its task is to copy the sections created by linker into the process memory map. To complete its tasks, it does not need to know much about the inner structure of the sections. Instead, all it worries about is whether the sections' attributes are read-only, read-write, and (as it will be discussed later) whether there needs to be some patching applied before the executable is ready for launching.

▪ **Note**   As will be shown later in discussions about the process of dynamic linking, the loader capabilities are a bit more complex than mere copying blocks of data.

It hence does not come as a big surprise that the loader tends to group the sections created by the linker into *segments* based on their common loading requirements. As shown in Figure 3-2, the loader segments typically carry several sections that have in common access attributes (read or read-and-write, or most importantly, to be patched or not).

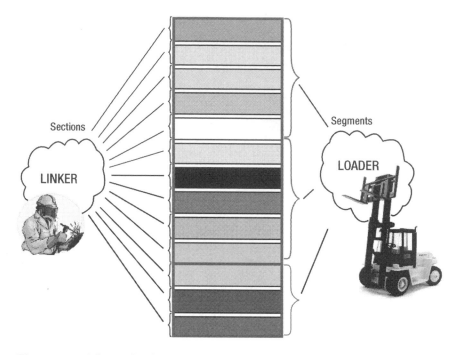

Figure 3-2. *Linker vs. loader*

As shown in Figure 3-3, using the readelf utility to examine the segments illustrates the grouping of many different linker sections into the loader segments.

```
milan@milan$ readelf --segments libmreloc.so

Elf file type is DYN (Shared object file)
Entry point 0x390
There are 7 program headers, starting at offset 52

Program Headers:
  Type           Offset   VirtAddr   PhysAddr   FileSiz MemSiz  Flg Align
  LOAD           0x000000 0x00000000 0x00000000 0x00540 0x00540 R E 0x1000
  LOAD           0x000f0c 0x00001f0c 0x00001f0c 0x00104 0x0010c RW  0x1000
  DYNAMIC        0x000f20 0x00001f20 0x00001f20 0x000c8 0x000c8 RW  0x4
  NOTE           0x000114 0x00000114 0x00000114 0x00024 0x00024 R   0x4
  GNU_EH_FRAME   0x0004c4 0x000004c4 0x000004c4 0x0001c 0x0001c R   0x4
  GNU_STACK      0x000000 0x00000000 0x00000000 0x00000 0x00000 RW  0x4
  GNU_RELRO      0x000f0c 0x00001f0c 0x00001f0c 0x000f4 0x000f4 R   0x1

 Section to Segment mapping:
  Segment Sections...
   00     .note.gnu.build-id .gnu.hash .dynsym .dynstr .gnu.version .gnu.version_r
          .rel.dyn .rel.plt .init .plt .text .fini .eh_frame_hdr .eh_frame
   01     .ctors .dtors .jcr .dynamic .got .got.plt .data .bss
   02     .dynamic
   03     .note.gnu.build-id
   04     .eh_frame_hdr
   05
   06     .ctors .dtors .jcr .dynamic .got
milan@milan$
```

Figure 3-3. *Sections grouped into segments*

Program Loading Stage

Once the binary format is identified, the role of the kernel's loader module comes to play. The loader first tries to locate the PT_INTERP segment in the executable binary file, which will assist it in the dynamic loading task.

In order to avoid the pitfalls of the proverbial "cart ahead of the horse" situation—since the dynamic loading is yet to be explained—let's assume the simplest possible scenario in which the program is statically linked and there is no need for dynamic loading of any kind.

STATIC BUILD EXAMPLE

The term ***static build*** is used to indicate the executable, which does not have any of the dynamic linking dependencies whatsoever. All the external libraries needed for creating such executable are statically linked. As a consequence, the obtained binary is fully portable, as it does not require the presence of any system shared library (not even libc) in order to be executed. The benefit of full portability (which seldom requires such drastic measures) comes at the price of the greatly enlarged byte size of the executable.

Other than full portability, the reason for the statically built executable may be purely educational, as it lends itself well to the process of explaining the original, simplest possible roles of the loader.

The effect of static building may be illustrated with the example of plain and simple "Hello World" example. Let's use the same source file to build the two applications, one of which is built with the -static linker flag; see Listings 3-1 and 3-2.

Listing 3-1. main.cpp

```
#include <stdio.h>

int main(int argc, char* argv[])
{
    printf("Hello, world\n");
    return 0;
}
```

Listing 3-2. build.sh

```
gcc main.cpp -o regularBuild
gcc -static main.cpp  -o staticBuild
```

The comparison of the byte sizes of the two executables will show that the byte size of the executable built statically is much larger (about 100 times in this particular example).

The loader continues by reading in the program's binary file segments' headers to determine the addresses and byte lengths of each of the segments. An important detail to point out is that at this stage the loader still does not write anything to the program memory map. All the loader does at this stage is to establish and maintain a set of structures (vm_are_struct for example), carrying the mappings between the segments of executable file (actually page-wide parts of each of the segments) and the program memory map.

The actual copying of segments from the executable happens after the program execution starts. By that time the virtual memory mapping between the page of physical memory granted to the process and the program memory map has been established; the first paging requests start arriving from the kernel requesting that page-wide ranges of program segments be available for execution. As a direct consequence of such policy, only the parts of the program that are actually needed at runtime happen to get loaded (Figure 3-4).

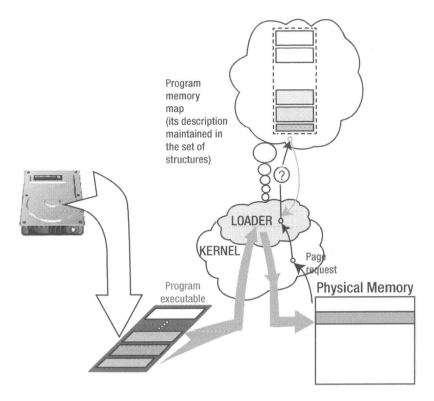

Figure 3-4. *Program loading stage*

Executing Program Entry Point

From the usual C/C++ programming perspective, the program entry point is the main() function. From the point of program execution, however, it is not. Prior to the point when the execution flow reaches the main() function, a few other functions get executed, which level the playfield for the program to run.

Let's take a closer look at what typically happens in Linux between the program loading and the execution of the main() function's first line of code.

The Loader Finds the Entry Point

After loading the program (i.e. preparing the program blueprint and copying the necessary sections to the memory for its execution), the loader takes a quick look at the value of e_entry field from the ELF header. This value contains the program memory address from which the execution will start.

Disassembling the executable binary file typically shows that the e_entry value carries nothing less than the first address of the code (.text) section. Coincidentally, this program memory address typically denotes the origin of the _start function.

The following is the disassembly of section .text:

```
08048320 <_start>:
 8048320:        31 ed                   xor     ebp,ebp
 8048322:        5e                      pop     esi
 8048323:        89 e1                   mov     ecx,esp
 8048325:        83 e4 f0                and     esp,0xfffffff0
 8048328:        50                      push    eax
 8048329:        54                      push    esp
 804832a:        52                      push    edx
 804832b:        68 60 84 04 08          push    0x8048460
 8048330:        68 f0 83 04 08          push    0x80483f0
 8048335:        51                      push    ecx
 8048336:        56                      push    esi
 8048337:        68 d4 83 04 08          push    0x80483d4
 804833c:        e8 cf ff ff ff          call    8048310 <__libc_start_main@plt>
 8048341:        f4                      hlt
```

The Role of _start( ) Function

The role of the _start function is to prepare the input arguments for the __libc_start_main function that will be called next. Its prototype is defined as

```
int __libc_start_main(int (*main) (int, char * *, char * *), /* address of main function    */
                      int argc,                 /* number of command line args     */
                      char * * ubp_av,          /* command line arg array          */
                      void (*init) (void),      /* address of init function        */
                      void (*fini) (void),      /* address of fini function        */
                      void (*rtld_fini) (void), /* address of dynamic linker fini function */
                      void (* stack_end)        /* end of the stack address        */
                     );
```

In fact, all that the instructions prior to the *call* instruction do is to stack up the required arguments for the call in the expected order.

In order to understand what exactly these instructions do and why, please take a look at the next section, which is dedicated to explaining the stack mechanism. But before going there, let's first complete the story about starting the program execution.

The Role of __libc_start_main( ) Function

This function is the key player in the process of preparing the environment for the program to run. It not only sets up the environment variables for the program during the program execution, but it also does the following:

- Starts up the program's threading.

- Calls the _init() function, which performs initializations required to be completed before the main() function starts.

  The GCC compiler through the __attribute__ ((constructor)) keyword supports custom design of the routines you may want to be completed before your program starts.

- Registers the _fini() and _rtld_fini() functions to be called to cleanup after the program terminates. Typically, the action of _fini() is inverse to the actions of the _init() function.

  The GCC compiler, through the __attribute__ ((destructor)) keyword, supports custom design of the routines you may want to complete before your program starts.

Finally, after all the prerequisite actions have been completed, the __libc_start_main() calls the main() function, thus making your program run.

Stack and Calling Conventions

As anyone with programming experience above the absolute beginner level knows, the typical program flow is in fact a sequence of function calls. Typically, the main function calls at least one function, which in turn may call huge number of other functions.

The concept of stack is the cornerstone of the mechanism of function calls. This particular aspect of program execution is not of paramount importance for the overall topic of this book, and we will not spend much more time discussing the details of how stack works. This topic has been a commonplace one for a long time, and there is no need to reiterate the well-known facts.

Instead, only the few important points related to the stack and functions will be pointed out.

- The process memory map reserves certain area for the needs of the stack.

- The amount of stack memory used at runtime actually varies; the larger the sequence of function calls, the more of stack memory is in use.

- The stack memory is not unlimited. Instead, the amount of available stack memory is bound with the amount of memory available for allocation (which is the part of the process memory known as *heap*).

Functions Calling Conventions

How a function passes the arguments to the function it calls is very interesting topic. A variety of very elaborate mechanisms of passing the variables to the functions have been designed, resulting in specific assembly language routines. Such stack implementation mechanisms are typically referred to as ***calling conventions***.

As a matter of fact, many different calling conventions have been developed for X86 architecture, such as cdecl, stdcall, fastcall, thiscall, to name just a few. Each of them is tailored for a specific scenario from a variety of design standpoints. The article titled *"Calling Conventions Demystified"* by Nemanja Trifunovic (www.codeproject.com/Articles/1388/Calling-Conventions-Demystified) provides an interesting insight into the differences between various calling conventions. The legendary Raymond Chen's series of blog articles titled "The History of Calling Conventions," which came a few years later (http://blogs.msdn.com/b/oldnewthing/archive/2004/01/02/47184.aspx), are probably the most complete single source of information about the topic.

Without spending too much time on this particular topic, a detail of particular importance is that among all the available calling conventions, one of them in particular, the cdecl calling convention, is strongly preferred for implementing the interface of dynamic libraries exported to the other world. Stay tuned for more details, as the discussions in Chapter 6 about library ABI functions will provide better insight into the topic.

CHAPTER 4

■ ■ ■

The Impact of Reusing Concept

The code reuse concept is omnipresent and has found an impressive variety of ways of manifesting itself. Its impact on the process of building programs happened much before the well-known transition from procedural programming languages toward the object-oriented ones.

The initial reasons for dividing the tasks between compiler and linker have been already described in the previous chapters. Briefly, it all started from the useful habit of keeping code in separate source files; then, at compile time it became obvious that compiler could not easily complete the task of resolving the references simply because tiling the code sections into the ultimate puzzle of program memory map had to happen first.

The idea of code reuse added the extra argument to the decision of splitting the compiling and linking stages. The amount of indeterminism brought in by the object files (all sections having zero-based address ranges plus unresolved references), which initially certainly looked like a drawback, in light of the code sharing idea actually started looking like a precious new quality.

The code reuse concept applied to the domain of building the program executables found its first realization in the form of **static libraries**, which are bundled collections of object files. Later on, with the advent of multitasking operating systems, another form of reusing called **dynamic libraries** came to prominence. Nowadays both concepts (static as well as dynamic libraries) are in use, each having pros and cons, hence requiring a deeper understanding of their functionality's inner details. This chapter describes in great detail these two somewhat similar, but also substantially different, concepts.

Static Libraries

The idea behind the concept of static libraries is exceptionally simple: once the compiler translates a collection of translation units (i.e., source files) into the binary object files, you may want to keep the object files for later use in other projects, where they may be readily combined at link time with the object files indigenous to that other project.

In order for it to be possible to integrate the binary object files into some other project, at least one additional requirement needs to be satisfied: that the binary files be accompanied by the export header include file, which will provide the variety of definitions and function declarations of at least these functions that can be used as entry points. The section titled "The Conclusion: The Impact of the Binary Reuse Concept" explains why some functions are more important than others.

There are several ways a set of object files may be put to use in various projects:

- The trivial solution is to save the object files generated by the compiler, and copy (cut-and-paste) or transfer in any way possible to a project that needs them (where they will be linked alongside the other object files into the executable), as shown in Figure 4-1.

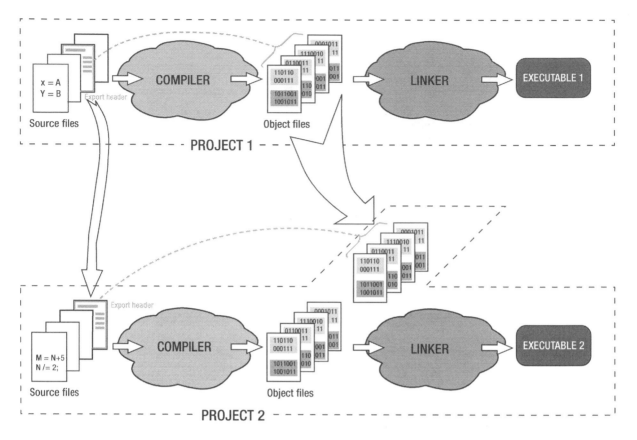

Figure 4-1. *A trivial method of binary code reuse, the precursor to static libraries*

- The better way is to bundle the object files into a single binary file, a static library.
  It is far simpler and far more elegant to deliver a single binary file to the other project than
  each object file separately (Figure 4-2).

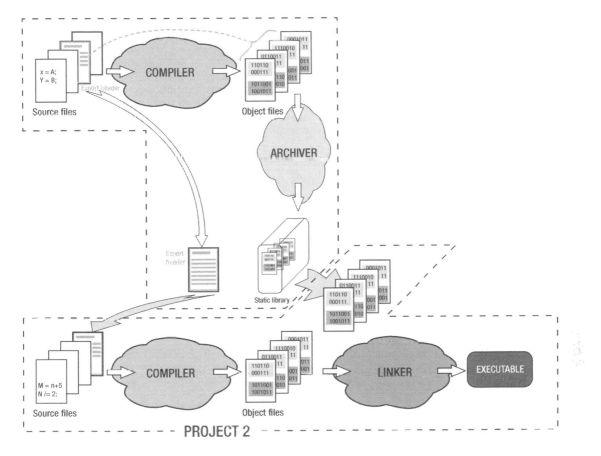

Figure 4-2. *The static library as form of binary code reuse*

- The obvious requirement in this case is that linker understands the static library file format and is capable of extracting its contents (i.e., object files bundled together) in order to link them in. Fortunately, this requirement has been met by probably each and every linker since the early days of microprocessor programming.

- Let it be noted also that the process of creating a static library is by no means irreversible. More specifically, a static library is merely an archive of object files, which can be manipulated in a number of ways. Through the easy use of appropriate tools, a static library can be dismantled into the collection of original object files; one or more object files may be thrown out from the library, new object files can be added, and finally, the existing object files may be replaced by a newer version.

Whichever of these two approaches you decide to follow, the trivial one or the more sophisticated static library approach, you essentially have the process of binary code reuse happening, as the binary files generated in one project are used in the other projects. The overall impact of binary code reuse to the landscape of software design will be discussed in detail later.

Dynamic Libraries

Unlike the concept of static libraries, which has been around since the early days of assembler programming, the concept of dynamic libraries came to full acceptance much later. The circumstances that led to its creation and adoption are tightly related to the appearance of multitasking operating systems.

In any analysis of the functioning of a multitasking operating system, one particular notion quickly comes to prominence: regardless of the variety of concurrent tasks, certain system resources are unique and must be shared by everybody. The typical examples of shared resources on the desktop system are the keyboard, the mouse, the video graphics adapter, the sound card, the network card, and so on.

It would be counterproductive and even disastrous if each and every application that intends to access the common resources had to incorporate the code (either as a source or as a static library) that provides control over the resource. This would be greatly inefficient, clumsy, and a lot of storage (both hard disk and memory) would be wasted on storing the duplicates of the same code.

The day-dreaming of better and more efficient operating systems led to the idea of having a sharing mechanism that would assume neither compiling in the duplicate source files nor linking in the duplicate object files. Instead, it would be implemented as some kind of runtime sharing. In other words, the running application would be capable of integrating in its program memory map the compiled and linked parts of some other executable, where the integration would happen on-demand, as-per-need, at runtime. This concept is referred to as dynamic linking/dynamic loading, which will be illustrated in more detail in the next section.

From the very early design stages one important fact became obvious: *of all the parts of the dynamic library, it only makes sense to share its code (.text) section, but not the data* with the other processes. In the culinary analogy, a bunch of different chefs can share the same cookbook (code). However, given that different chefs may be concurrently preparing utterly different dishes from the same cookbook, it would be disastrous if they shared the same kitchen utensils (data).

Obviously, if a bunch of different processes had access to the dynamic library data section, the variable overwriting would happen at arbitrary moments, and the execution of the dynamic library would be unpredictable, which would render the whole idea meaningless. This way, by mapping only the code section, the multiple applications are free to run the shared code each in its own compartment, separately from each other.

Dynamic vs. Shared Libraries

The ambitions of operating system designers early on were to avoid unnecessary multiple presence of the same pieces of operating system code in the binaries of each and every application that may need them. For example, each application that needed to print the document would have to incorporate the complete printing stack, ending up with the printer driver, in order to provide the printing feature. If the printer driver changed, the whole army of application designers would need to recompile their applications; otherwise, a chaos would emanate due to the runtime presence of plethora of different printer driver versions.

Obviously, the right solution would be to implement the operating system in such a way that the following would happen:

- The commonly needed functionality is provided in the form of dynamic libraries.

- The application that needs the access at common functionality would need to merely load the dynamic library at runtime.

The basic idea behind the concept of dynamic libraries is illustrated in the Figure 4-3.

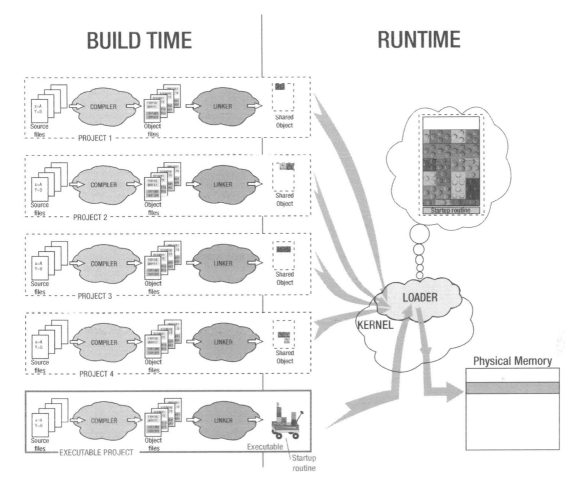

Figure 4-3. *The dynamic libraries concept*

The first solution to this problem (i.e., the first version of dynamic linking implementation, known as **load time relocation (LTR)**), achieved the goal with partial success. The good news was that applications were relieved of carrying the unnecessary baggage of operating system code in its binaries; instead, they were deployed with only the app-specific code, whereas all system-related needs were satisfied by dynamically linking the modules supplied by the operating system.

The bad news, however, was that if multiple applications needed certain system functionality at runtime, each of the applications had to load their own copy of the dynamic library. The underlying cause of this limitation was the fact that the load time relocation technique modified the symbols of the .text section of the dynamic library to fit the particular address mapping of the given application. For another application, which would load the dynamic library into the possibly different address range, the modified library code simply would not fit the different memory layout.

As a result, multiple copies of the dynamic libraries resided in the processes' memory maps at runtime. This is something that we could live with for some time, but the long term goals of the design were far more ambitious: to provide a more efficient mechanism that would allow the dynamic library to be loaded just once (by whatever application happens to load it first) and be made available to any other application that tried to load it next.

This goal was achieved through the concept known as **position independent code (PIC)**. By changing how the dynamic library code accessed the symbols, only one copy of the dynamic library loaded into the memory map of any process would become shareable by memory mapping it to any application's process memory map (Figure 4-4).

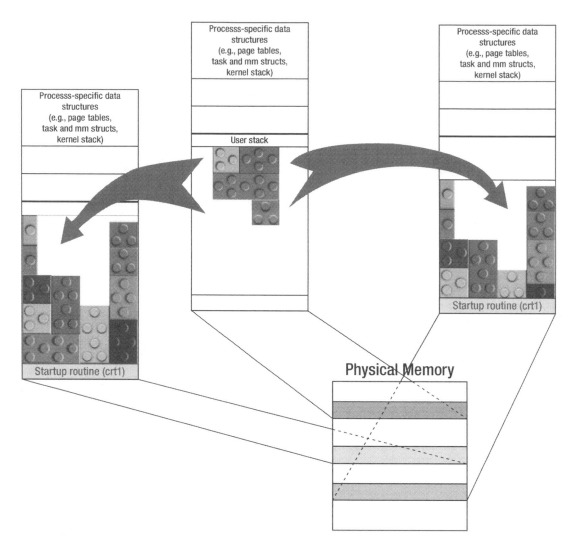

Figure 4-4. *The advances brought by the PIC technique of dynamic linking*

Furthermore, it is not unusual that the operating system loads certain common system resources (top level drivers, for example) into the physical memory, knowing that it will most likely be needed by the plethora of running processes. The effect of dynamic linking is that each of the processes has the perfect illusion that they are the sole owners of the driver.

Since the invention of PIC concept, the dynamic libraries designed to support it were called shared libraries. Nowadays, the PIC concept is prevalent, and on 64-bit systems it is strongly favored by the compilers, so the naming distinction between the terms dynamic vs. shared library is disappearing, and the two names are used more or less interchangeably.

The concept of virtual memory paved the foundation of the success of the idea of runtime sharing (epitomized by the concept of position-independent code). The initial idea is fairly simple: if the real process memory map (with real, concrete addresses) is nothing but the result of 1:1 mapping of the zero-based process memory map, what really prevents us from creating a monster, a real process memory map obtained by mapping parts of more than one different processes? In fact, this is exactly how the mechanism of runtime sharing of dynamic libraries works.

The successful implementation of the PIC concept represents a cornerstone of modern multitasking operating systems.

Dynamic Linking in More Detail

The concept of dynamic linking is at the very core of the concept of dynamic libraries. It is practically impossible to fully understand how the dynamic libraries work without understanding the complex interplay between the dynamic library, the client executable, and the operating system. The focus of this section is to provide the necessary broad level understanding of the process of dynamic linking. Once its essence is understood, the subsequent sections of this document will pay the due attention to the details.

So, let's see what really happens during the process of dynamic linking.

Part 1: Building the Dynamic Library

As the previous figures suggest, the process of building a dynamic library is a complete build, as it encompasses both compilation (converting the source into the binary object files) as well as resolving the references. The product of the dynamic library building process is the binary file whose nature is identical to the nature of executable, the only difference being that dynamic library lacks the startup routines that would allow it to be started as independent program (Figure 4-5).

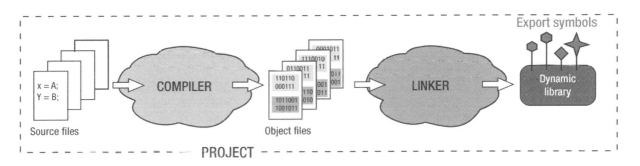

Figure 4-5. *Building the dynamic library*

Here are some notes to consider:

- In Windows, building a dynamic library strictly requires that all the references must be resolved. If the dynamic library code calls a function in some other dynamic library, that other library and the references symbol it contains must be known at build time.

- In Linux, however, the default option allows some more flexibility, allowing that some the symbols be unresolved with the expectation that they will eventually show up in the final binary after some other dynamic library is linked in. Additionally, the Linux linker provides the option to fully match the Windows linker's strictness.

- In Linux, it is possible to modify the dynamic library to make it runnable by itself (still researching whether such option exists on Windows). In fact, the libc (C runtime library) is executable by itself; when invoked by typing its filename into the shell window, it prints a message on the screen and terminates. For more details of how to implement such feature, please check Chapter 14.

Part 2: Playing by Trust While Building the Client Executable (Looking for the Symbols Only)

The next stage in the scenario of using a dynamic library happens when you try to build the executable that intends on using the dynamic library at runtime. Unlike the static libraries scenario in which the linker is creating the executable on its own at will, the scenario of linking the dynamic libraries is peculiar in that the linker tries to combine its current work with the existing results of the previously completed linking procedure that created dynamic library binary.

The crucial detail in this part of the story is that linker pays almost all of its attention to the dynamic library's symbols. It appears that at this stage the linker is almost not interested in any of the sections, neither code (.text), nor data (.data/.bss).

More specifically, the linker in this stage of operation "plays it by trust."

It does not examine the binary of the dynamic library thoroughly; it neither tries to find the sections or their sizes, nor attempts to integrate them into the resultant binary. Instead, it solely tries to verify that the dynamic library contains the symbols needed by the resultant binary. Once it finds it, it completes the task and creates the executable binary (see Figure 4-6).

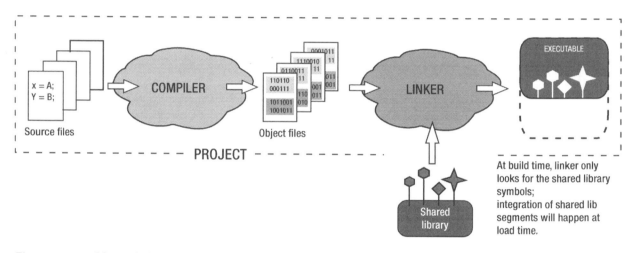

Figure 4-6. Build time linking with a dynamic library

The approach of "playing by trust" is not completely unintuitive. Let's consider a real-life example: if you tell someone that in order to mail a letter he needs to go to the kiosk in the nearby square and buy a postage stamp, you are essentially basing your advice on reasonable amount of trust. You do know that there should be a kiosk on the square, and that it carries postage stamps. The fact that you don't know the particular details of the kiosk operation (working hours, who works there, the price of postage stamps) does not diminish the validity of your advice, as at runtime all these less important details will be resolved. The idea of dynamic linking is based on completely analogous assumptions.

Let it be noticed, however, that this amount of trust leaves open doors for many interesting scenarios, all of which fall under the "build with one, load the other" paradigm. The practical implications vary from peculiar software design tricks all the way to the whole new paradigm (plug-ins), both of which will be discussed later in the book.

Part 3: Runtime Loading and Symbol Resolution

The events happening at load time are of crucial importance, as this is the time when the confidence that the linker had in the dynamic library's promises needs to be confirmed. Previously, the build procedure (possibly completed on a build machine "A") examined the copy of dynamic library binary in the search for symbols that executable needs. Now what needs to happen at runtime (possibly on different, runtime machine "B") is the following:

1. The dynamic library binary file needs to be found.

    Each operating system has a set of rules stipulating in which directory the loader should look for the dynamic libraries' binaries

2. The dynamic library needs to be successfully loaded into the process.

    This is the moment where the promise of build-time linking must be fulfilled at runtime.

    In fact, the dynamic library loaded at runtime *must carry the identical set of symbols promised to be available at build time*. More specifically, in the case of function symbols the term "identical" means that the function symbols found in the dynamic library at runtime must exactly match the complete function signature (affiliations, name, list of arguments, linkage/calling convention) promised at build time.

    Interestingly enough, it is **not** required that the actual assembly code (i.e., the sections contents) of the dynamic library found at runtime matches the code found in the dynamic library binary used during the build time. This opens up a lot of interesting scenarios which will be discussed in detail later on.

3. The executable symbols need to be resolved to point to the right address in the part of process of memory map where the dynamic library is mapped into.

    It is this stage the integration of dynamic library into the process memory map truly deserves to be called **dynamic linking**, as unlike the conventional linking it happens at load time.

If all the steps of this stage completed successfully, you may have your application executing the code contained in the dynamic library, as illustrated in Figure 4-7.

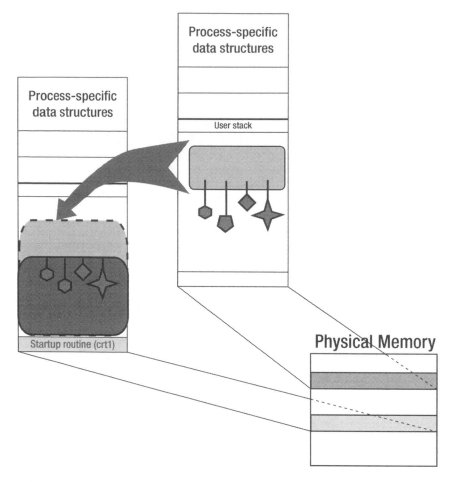

Figure 4-7. *Load time linking of a dynamic library*

Peculiarities of Dynamic Linking on Windows

As it is true that the dynamic linking happens in two phases (build time vs. runtime) in which the linker focuses on different details of dynamic library binary, there is no substantial reason as to why the same identical copy of the dynamic library binary couldn't be used in both of the phases.

Even though in the build-time dynamic linking phase only the library symbols play a role, there is nothing wrong if the exact same copy of the binary file is used in the runtime phase as well.

This principle is mostly followed throughout the variety of operating systems, including Linux. In Windows, however, in the attempt to make the separation between the dynamic linking stages clearer, things are made slightly more complicated in a way that can confuse beginners a bit.

Special Binary File Types Related to Dynamic Linking in Windows

In Windows, the distinction between the different phases of dynamic linking is accentuated by using slightly different binary file types in each of the phases. Namely, when a Windows DLL project is created and built, the compiler produces several different files.

Dynamically Linked Library (.dll)

This file type is in fact the dynamic library, a shared object used at runtime by the processes through the mechanism of dynamic linking. More specifically, the majority of facts presented so far about the principles on which the dynamic library functions are fully applicable to the DLL files.

Import Library File (.lib)

A dedicated **import library (.lib)** binary file is used on Windows specifically at the "Part2" phase of dynamic linking (Figure 4-8). It contains only the list of DLL symbols and none of its linker sections, and its purpose is solely to present the set of dynamic library's exported symbols to the client binary.

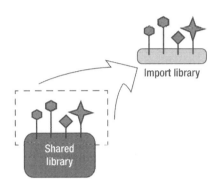

Figure 4-8. *Windows import library*

The file extension of the import library file (.lib) is the potential source of confusion, as the same file extension is also used to indicate the static libraries.

Another detail that deserves a bit of discussion is the fact that this file is called *import library* but in fact plays a role in the process of *exporting* the DLL symbols. As it is true that the choice of naming does depend on the side from which we look at the process of dynamic linking, it is also true that this file belongs to the DLL project, gets created by building DLL project, and may be disseminated to uncountable many applications. For all of these reasons, it should not be wrong to adopt the "from DLL outward" direction, and hence use the name *export library*.

The obvious proof that someone else at Microsoft shared this viewpoint at least to an extent can be found in the section discussing the use of __declspec keyword, where the naming (__declspec(dllexport)) is used to indicate export from DLL toward the client apps (i.e., in the outward direction).

One of the reasons why folks at Microsoft decided to stick with this particular naming convention is that the DLL project produces another type of library file that may be used instead of this one in the scenarios of circular dependencies. That another file type is called an **export file (.exp)** (see below), and in order to distinguish between the two, the existing naming has been retained.

Export File (.exp)

The *export file* has the same nature as the *import library* file. However, it is typically used in the scenario when two executables have circular dependencies that make impossible to complete the building of either one. In such case, the exp file is provided with the intention to make it possible for at least one of the binaries to successfully compile, which in turn can be used by the other dependent binaries to complete their builds.

■ **Note**   Windows DLLs are strictly required to resolve all the symbols at build time. On Linux, however, it is possible to leave some dynamic library symbols unresolved with the expectation that missing symbols will eventually appear in the process memory map as a result of dynamically linking in some other dynamic libraries.

Unique Nature of Dynamic Library

It is important to understand early on that in the ensemble of binary types the dynamic library has fairly unique nature, the details of which are important to keep in mind when dealing with the usual related design issues.

When looking at the other binary types, the opposite natures of the executables and the static libraries become obvious almost immediately. The creation of a static library does not involve the linking stage, whereas in the case of the executable it is the mandatory last step. As a consequence, the nature of the executable is far more completed, as it contains resolved references, and due to the embedded extra start routines, it is ready for execution.

In that regard, despite the word "library," which suggests similarities between the static and dynamic libraries, it is the fact that the nature of the dynamic library is far closer to the nature of the executable.

Property 1: Dynamic Library Creation Requires the Complete Build Procedure

The process of creating the dynamic library involves not only compiling but the linking stage as well. Despite what the naming similarities suggest, the completeness of the dynamic library build process (i.e., linking in addition to compiling) makes the dynamic library far more similar to the executable than to the static library. The only difference is that executable contains the startup code that allows the kernel to start the process. It is definitely possible (in Linux for sure) to add a few lines of code to the dynamic library that make it possible to execute the library from the command line as if it were an executable binary type. For more details, please check Chapter 14.

Property 2: The Dynamic Library Can Link In Other Libraries

This is a really interesting fact: it is not only the executable that can load and link the dynamic library, but it can be also another dynamic library. Hence, we can no longer say "executable" to indicate the binary that links in a dynamic library; we must use other more appropriate term.

■ **Note**   I decided to hereafter use the term **"client binary"** to indicate the executable or the dynamic library that loads a dynamic library.

Application Binary Interface (ABI)

When the interfacing concept is applied to the domain of programming languages, it is typically used to denote the structure of function pointers. C++ adds a few extra meanings by defining it as a class of function pointers; additionally, by declaring the function pointers to be equal to NULL, the interface gets an extra kick of abstraction as it becomes unsuitable for instantiation, but can be used as idealistic model for other classes to implement it.

The interface exported by a software module to the clients is typically referred to as an **application programming interface (API)**. When applied to the domain of binaries, the concept of an interface gets one additional domain-specific flavor called an **application binary interface (ABI)**. It is not wrong to think of ABI as a set of symbols (primarily a set of function entry points) created in the process of compiling/linking of the source code interface.

The ABI concept comes in handy when explaining more precisely what happens during the dynamic linking.

- During the first (build-time) phase of the dynamic linking, the client binary in fact links against the library's exported ABI.

  As I pointed out, at build time the client binary in fact only checks whether the dynamic library exports the symbols (function pointers such as the ABI) and does not care at all about the sections (the function bodies).

- In order to successfully complete the second (runtime) phase of dynamic linking, the binary specimen of the dynamic library available at runtime *must export the unchanged ABI, identical to that found at build time.*

The second statement is considered as *the basic requirement of dynamic linking.*

Static vs. Dynamic Libraries Comparison Points

Even though I just barely touched on the concepts behind the static and dynamic libraries, some comparisons between the two can be already drawn.

Differences in Import Selectiveness Criteria

The most interesting difference between the static and dynamic libraries is the difference in selectiveness criteria applied by a client binary that tries to link them.

Import Selectiveness Criteria for Static Libraries

When the client binary links the static library, it does not link in the complete static library contents. Instead, it links in strictly and solely only the object files containing the symbols that are really needed, as shown in Figure 4-9.

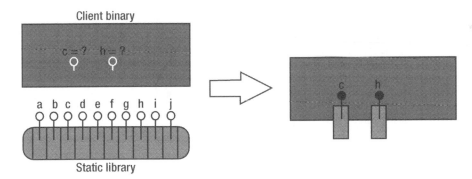

Figure 4-9. *Import selectiveness criteria for static libraries*

The byte length of the client binary gets increased, albeit only by the amount of relevant code ingested from the static library.

■ **Note** Despite the fact that the linking algorithm is selective when choosing which object files to link in, the selectivity does not go beyond the granularity of individual object files. It still could happen that in addition to the symbols that are really needed, the chosen object file contains some symbols that are not needed.

Import Selectiveness Criteria for Dynamic Libraries

When the client binary links the dynamic library, it features the selectivity only at the level of the symbol table, in which only the dynamic library symbols that are really needed become mentioned in the symbol table.

In all other regards, the selectivity is practically nonexistent. Regardless of how small a portion of the dynamic library functionality is concretely needed, the entire dynamic library gets dynamically linked in (Figure 4-10).

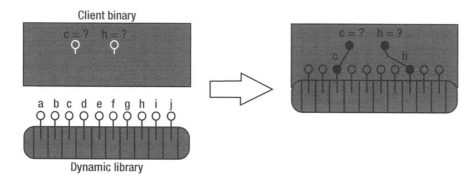

Figure 4-10. *Import selectiveness criteria for dynamic libraries*

The increased amount of code only happens at runtime. The byte length of the client binary does not get increased significantly. The extra bytes needed for the bookkeeping of new symbols tend to amount to small byte counts. However, linking the dynamic library imposes the requirement that the dynamic library binary will need to be available at runtime on the target machine.

Whole Archive Import Scenario

An interesting twist happens when the static library's functionality needs to be presented to the binary clients through the intermediary dynamic library (Figure 4-11).

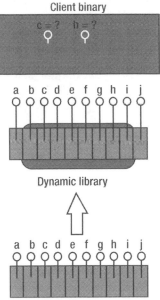

Figure 4-11. *"Whole archive" scenario of importing static library*

The intermediary dynamic library itself does not need any of the static library's functionality. Therefore, according to the formulated import selectiveness rules, it will not link in anything from the static library. Yet, the sole reason why the dynamic library is designed is to ingest the static library functionality and export its symbols for the rest of the world to use.

How to mitigate these opposite requirements?

Fortunately, this scenario has been early identified, and adequate linker support has been provided through the --whole-archive linker flag. When specified, this linker flag indicates that one or more libraries listed thereafter will be unconditionally linked in entirely, regardless of whether the client binary that links them needs their symbols or not.

In recognition of this scenario, the Android native development system in addition to supporting the LOCAL_STATIC_LIBRARIES build variable, the native build system also supports the LOCAL_WHOLE_STATIC_LIBRARIES build variable, like so:

```
$ gcc -fPIC <source files> -Wl,--whole-archive -l<static libraries> -o <shlib filename>
```

Interestingly, there is a counter-action linker flag (--no-whole-archive). Its effect is to counteract the effect of --whole-archive for all subsequent libraries being specified for linking on the very same linker command line.

```
$ gcc -fPIC <source files> -o <executable-output-file> \
    -Wl,--whole-archive -l<libraries-to-be-entirely-linked-in> \
    -Wl,--no-whole-archive -l<all-other-libraries>
```

Somewhat similar in nature to the --whole-archive flag is the -rdynamic linker flag. By passing this linker flag you are basically requesting that the linker exports all the symbols (present in the .symtab section) to the dynamic (.dynsym) section, which basically makes them usable for the purposes of dynamic linking. Interestingly, this flag does not seem to require the -Wl prefix.

Deployment Dilemma Scenarios

When designing the software deployment packages, the build engineers typically face a requirement to minimize the byte size of the deployment package. In one of the simplest possible scenarios, the software product that needs to be deployed is comprised of an executable that delegates the task of providing certain part of its functionality to a library. Let's say that the library may come in both flavors, the static as well as the dynamic library. The basic question that the build engineer faces is which kind of linking scenarios to utilize in order to minimize the byte size of the deployed software package.

Choice 1: Linking with a Static Library

One of the choices that a build engineer faces is to link the executable with the static version of the library. This decision comes with pros and cons.

- **Pros:** The executable is completely self-contained, as it carries all the code it needs.

- **Cons:** The executable byte size gets increased by the amount of code ingested from the static library.

Choice 2: Linking with a Dynamic Library

Another possibility, of course, is to link the executable with the dynamic version of the library. This decision also comes with pros and cons.

- **Pros:** The executable byte size does not get changed (except maybe by the small symbols bookkeeping expense).

- **Cons:** There is always a chance that the required dynamic library for whatever reason is not physically available on the target machine. If precaution is taken and the required dynamic library gets deployed together with the executable, several potential problems may ensue.

  - First, the overall byte size of the deployment package definitely gets larger, as you now deploy an executable and a dynamic library.

  - Second, the deployed dynamic library version may not match the requirements of the other applications that may rely on it.

  - Third, fourth, and so on, there is a whole set of problems that may happen when dealing with the dynamic libraries, known under the name "DLL hell."

Final Verdict

The linking with static libraries is a good choice when the application links in relatively smaller portions of relatively smaller number of static libraries.

The linking with dynamic libraries comes as a good choice when the application depends on the dynamic libraries expected with the great certainty to exist at runtime on the target machine.

The likely candidates are OS-specific dynamic libraries, such as C runtime library, graphic subsystems, user space top level device drivers, and/or the libraries coming from very popular software packages. Table 4-1 summarizes the differences between dealing with static vs. dynamic libraries.

Table 4-1. *Comparison Points Summary*

Comparison Category	Static Libraries	Dynamic Libraries
Build Procedure	Incomplete: Compiling: yesLinking: **no**	Complete: Compiling: yesLinking: **yes**
Nature of Binary	Archive of **object file**(s) All the sections exist, but the majority of references are unresolved (except local references). Can't exist standalone; the circumstances of the clientbinary determine plenty of details. All of its symbolshave some meaning only within the client executable.	The **executable** without the startup routines. Contains resolved references (except when intended otherwise), some of which are intended to be globally visible. Very independent (in Linux, with a few simple additions the missing startup routines can be effectively added). Highly specialized in certain strategic tasks; once loaded into the process, typically very dependable and reliable in providing specialized services.
Integration With Executable	Happens during the executable building process,completed during the linking stage. **Efficient: Only the needed object files from the archive get linked into the executable.** **The byte size of the client binary gets increased, though.**	Happens through two separate phases of dynamic linking: 1) Linking against the available symbols 2) Symbols and sections integration at load time **Inefficient: The complete library gets loaded into the process regardless of which part of library is really needed.** **The byte size of client binary almost does not change. However, the availability of the dynamic library binary at runtime is one extra thing to worry about.**
Impact on Executable Size	Increases the executable size, as sections get addedto the executable sections.	Reduces the executable size, as only the app-specific code resides in the app executable, whereas the shareable parts are extracted into the dynamic library.
Portability	Excellent, as everything the app needs is within its binary. The absence of external dependencies makes the portability easy.	Varies. Good for OS-standard dynamic libraries (libc, device drivers, etc.), as they are guaranteed to exist on runtime machine. Less good for app-specific or vendor-specific scenarios. Plenty of scenarios for potential problems exist (versioning, missing libraries, search paths, etc.)

(continued)

Table 4-1. (*continued*)

Comparison Category	Static Libraries	Dynamic Libraries
Ease of Combining	Very limited.	Excellent.
	Can't create a static library by using the other libraries (neither static nor dynamic). Can only link all of them together into the same executable.	A dynamic library can link in one or more static libraries, and/or one of more dynamic libraries.
		In fact, Linux can be viewed as "Legoland," a set of constructions made by dynamic libraries linking with the other dynamic libraries. The magnitude of integration is greatly facilitated by the availability of source code.
Ease of Converting	Fairly easy.	Practically impossible for most mortals.
	The standard function of the archiver utility is the extraction of ingredient object files. Once extracted, they can be eliminated, replaced, or recombined into a new static or dynamic library.	Some commercial solutions have been seen that attempt with various degree of success to implement the conversion of a dynamic to a static library.
	Only in very exceptionally special cases (in the "Tips and Tricks" section, see the topic about linking the static lib into a dynamic lib on 64-bit Linux) this may not be good enough, and you might need to recompile the original sources.	
Suitable for Development	Cumbersome.	Excellent.
	Even the smallest changes in the code require recompiling all executables that link the library.	The best way to work on an isolated feature is to extract it into the dynamic library.
		As long as the exported symbols (function signatures, and/or data structure layouts) are not changed, recompiling the library does not require recompiling the rest of the code.
Misc/Other	Simpler, older, ubiquitous form of binary sharing applied even in the simplest microcontroller developmentenvironments.	Newer way of binary code reuse.
		Modern multitasking system can't even be imagined without them.
		Essential to the concept of plug-ins.

Useful Comparison Analogies

Tables 4-2 through 4-4 list several very useful and illustrative analogies, which may help you better understand the role of compilation process.

Table 4-2. *Legal Analogy*

Binary Type	Legal Equivalent
Static Library	**The law paragraph**
	In general, it is written in a kind of indeterministic fashion. For example: If a person (*which person?*) is convicted of conducting a class A misdemeanor (*which particular misdemeanor? what exactly did the person do?*), he or she will be sentenced to pay the fine not exceeding 2000 dollars (*exactly how much?*), or to serve the prison term not exceeding 6 months (*exactly how long?*) or both (*which one of the three possible combinations?*).
Dynamic Library	**Concrete accusation**
	John Smith is convicted of resisting arrest and disobeying the police officer. The prosecution requests that he pay a fine of $1,500 and spend 30 days in jail.
Executable	**Serving the sentence**
	All references (who, what, when, and possibly why) are resolved: the law violations have been proven in the court of law, the judge sentenced John Smith according to the letter of law, and everything is ready for him to serve his sentence in the nearby state correction facility.

Table 4-3. *Culinary Analogy*

Binary Type	Culinary Equivalent
Static Library	**Raw food ingredients** (e.g., raw meat or raw vegetables)
	Definitely suitable for consumption, but can't be served right away, as they need a certain amount of processing (marinating, adding spices, combining with other ingredients and most importantly, termic processing) which must be completed first.
Dynamic Library	**Pre-cooked or ready-made dish**
	Ready for consumption, but serving it as-is makes very little sense. However, if the rest of the lunch is ready, it will make a great addition to the served meal.
Executable	**Complete lunch course**
	This consists of the fresh bread of the day, salad of the hour, and the prepared main course, which can be enriched by a certain amount of a warmed-up dish that was cooked a few days ago.

Table 4-4. *Tropical Jungle Expedition Analogy*

Binary Type	Expedition Role Equivalent
Executable	**British lord, the leader of the expedition** Decorated combat veteran, known for his excellent survival skills and instincts. Assigned by the British Geographic Society to investigate rumors that in the depths of tropical jungles exist the temples of long-lost advanced civilizations, hiding numerous material and scientific treasures. He is entitled to logistic support from the local British consular department, which takes care of coordinating the effort with the local authorities and provides all kinds of help with supplies, money, logistics, and transportation.
Dynamic Library	**Local hunter, the expedition guide** This guy was born and raised in the expedition target geographic area. He speaks all local languages, knows all tribal religions and cultures; has plenty of personal connections in the area; knows all dangerous places and how to avoid them; has exceptional survival skills; is a good hunter, excellent trail blazer, and can predict weather changes. Highly specialized in everything related to the jungle, and can completely take care of himself. Most of his adult time has been spent as a hired guide for expeditions like this one. Between expeditions he does pretty much nothing, other than spending his time with the family, going fishing and hunting, etc. Has neither the ambition nor the financial power to start anything himself.
Static Library	**Young personal assistant** Young British lad from the aristocratic family. Little or no real life experience, but Oxford degree in archeology and knowledge of ancient languages, as well as the operational knowledge of stenography, telegraphy, and Morse code earns him a place on the team. Even though his skills are potentially applicable to many roles and many scenarios, he has never been in the tropical areas, does not speak local languages, and will for the most part depend on higher authority and/or higher expertise of various kinds. Most likely he will have no formal authority over the course of expedition and no power of making the decisions of any kind except in the domain of his immediate expertise, and only when asked to do so.

■ **Note**   In the culinary analogy, you (the software designer) are running a restaurant in which (through the process of building the executables) you prepare a meal for the hungry CPU who is hardly waiting to start munching the meal.

The Conclusion: The Impact of Binary Reuse Concept

As soon as the concept of binary reuse was proven to work, it had the following immediate consequences to the landscape of software design:

- The appearance of dedicated projects whose intention is not to build executable code, but instead to build a binary bundle of reusable code.

- As soon as the practice of building the code for others to use starting gaining momentum, the necessity to follow the encapsulation principle came to prominence.

The essence of the idea of encapsulation is that if we are building something for others to use, it is always good if such export products come with clearly separated essential features from the less important inner functionality details. One of the mandatory ways to achieve it is to declare the interface, the set of quintessential functions that a user is most interested in using.

- The interface (set of quintessential/the most important functions) is typically declared in the export header file (an include file that provides the top-level interface between the reusable binary code and the potential user).

In a nutshell, the way to distribute the code for others to use is to deliver the software package carrying the set of binary files and the set of export header files. The binary files export the interface, mostly the set of functions quintessential for using the package.

The next wave of consequences followed immediately thereafter:

- The appearance of SDKs (software development kits), which are, in the most basic version, a set of export headers and binaries (static and/or dynamic libraries) intended for integration with binaries created while compiling source files indigenous to the client project.

- The appearance of the "one engine, variety of GUIs" paradigm.

  There are plenty of examples where the popular engine gets used by different applications presenting the different GUIs to the user, but running the same engine (loaded from the same dynamic libraries) in the background. Typical examples in the domain of multimedia are ffmpeg and avisynth.

- A potential for controlled exchange of intellectual properties.

  By delivering binaries instead of source code, the software companies may deliver their technology without disclosing the ideas behind it. The availability of disassemblers makes this story somewhat more complicated, but in the long run the basic idea still applies.

CHAPTER 5

■ ■ ■

Working with Static Libraries

In this chapter, I will review the typical life cycle in dealing with static libraries. I will start with simple guidelines for creating static library, then I will provide the overview of typical use case scenarios, and finally I will take a closer look at certain expert-level design tips and tricks.

Creating Static Library

A static library is created when the object files created by compiler from the set of source files are bundled together into a single archive file. This task is performed by a tool called an ***archiver***.

Creating Linux Static Library

On Linux, the archiver tool, called simply *ar*, is available as part of GCC toolchain. The following simple example demonstrates the process of creating static library out of two source files:

```
$ gcc -c first.c second.c
$ ar rcs libstaticlib.a first.o second.o
```

By Linux convention, static libraries names start with prefix *lib* and have the file extension *.a*.

In addition to performing its basic task of bundling the object files into the archive (static library), the ar can perform several additional tasks:

- Remove one or more object files from library.
- Replace one or more object files from the library.
- Extract one or more object files from the library.

The complete list of supported features can be found on the ar tool man page (http://linux.die.net/man/1/ar).

Creating a Windows Static Library

The task of creating the static library on Windows does not substantially differ from the same task performed in Linux. Even though it can be completed from the command line, the fact of life is that in the vast majority of cases the task of creating the static library is performed by creating a dedicated Visual Studio (or other similar IDE tool) project with the option of building the static library. When examining the project command line, you can see in Figure 5-1 that the task boils down to essentially the same use of an archiver tool (albeit a Windows version).

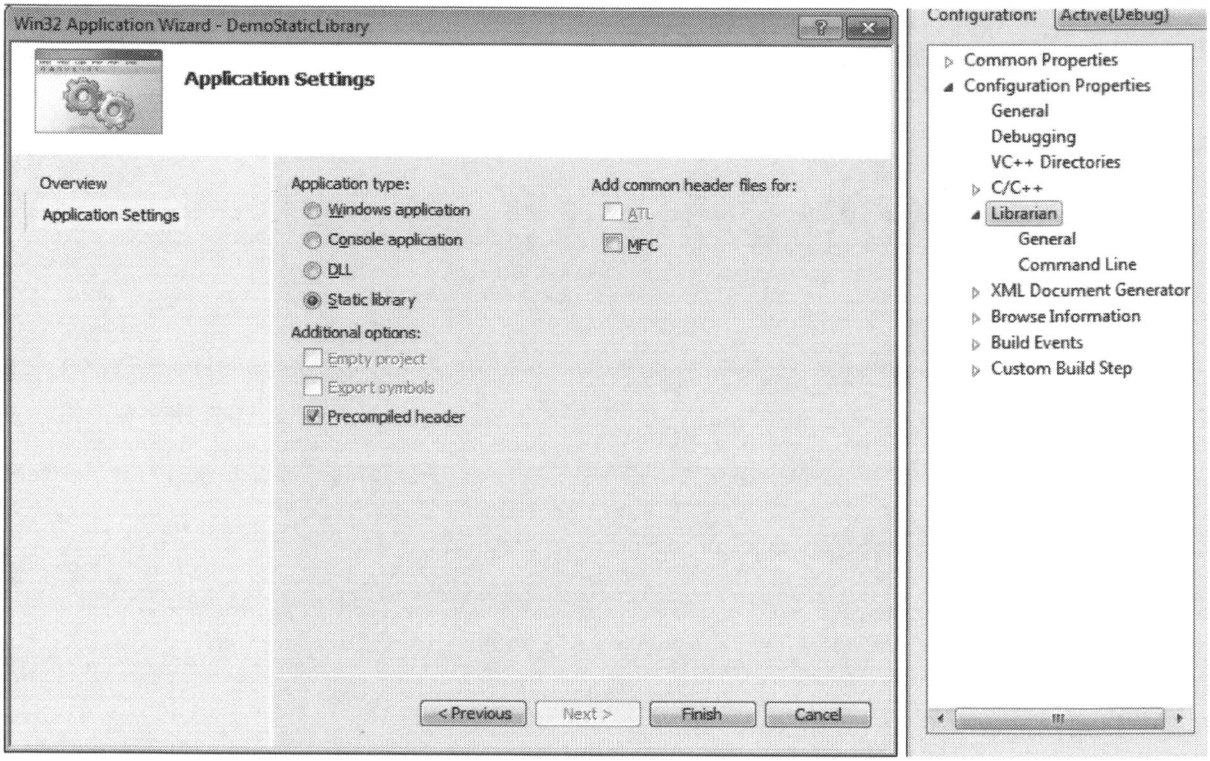

Figure 5-1. *Creating a Win32 static library*

Using the Static Library

Static libraries are used at the linking stage of projects that build executables or dynamic libraries. The static libraries' names are typically passed to the linker together with the list of object files that need to be linked in. If the project also links in the dynamic libraries, their names are the part of the same list of linker input arguments.

Recommended Use Case Scenarios

The static libraries are the most basic way of binary sharing the code, which has been available for a long time before the invention of dynamic libraries. In the meantime, the more sophisticated paradigm of dynamic libraries has taken over the domain of binary code sharing. However, there are still a few scenarios in which resorting the use of static libraries still makes sense.

The static libraries are perfectly suitable for all the scenarios implementing the core of various (mostly proprietary) algorithms, ranging from elementary algorithms such as search and sort all the way to very complex scientific or mathematical algorithms. The following factors can supply the extra push toward deciding to use the static library as the form of delivering the code:

- The overall code architecture can be described more as a "wide collection of various abilities" instead of a "module with the strictly defined interface."

- The actual computation does not rely on a specific OS resource (such as a graphics card's device driver, or high priority system timers, etc.) that requires loading the dynamic libraries.

- The end user wants to use your code, but does not necessarily want to share it with anybody else.

- Code deployment requirements suggest the need for monolithic deployment (i.e. small overall number of binary files delivered to the client's machine).

Using the static library always means tighter control over the code, albeit at the price of reduced flexibility. The modularity is typically reduced, and the appearance of new code versions typically means recompiling every application that uses it.

In the domain of multimedia, the signal processing (analysis, encoding, decoding, DSP) routines are typically delivered in the form of static libraries. On the other hand, their integration into the multimedia frameworks (DirectX, GStreamer, OpenMAX) are implemented in the form of dynamic libraries which link in the algorithm-related static libraries. In this scheme, the simple and strict duties of communicating with the framework are delegated to the thin shell of dynamic library part, whereas the signal processing complexities belong to the static library part.

Static Libraries Tips and Tricks

The following section covers the list of important tips and tricks related to the use of static libraries.

Potential for Losing the Symbol Visibility and Uniqueness

The way the linker integrates the static library sections and symbols into the client binary is genuinely fairly simple and straightforward. When linked into the client binary, the static library sections get seamlessly combined with the sections coming from the client binary's indigenous object files. The static library symbols become the part of the client binary symbols list and retain their original visibility; the static library's global symbols become the client binary's global symbols, and the static library's local symbols become the client binary's local symbols.

When the client binary is the dynamic library (i.e. not the application), the outcome of these simple and straightforward rules of integration may be compromised by the other dynamic libraries' design rules.

Where is the twist?

The implicit assumption in the concept of dynamic libraries is the modularity. It is not wrong to think of a dynamic library as of a module that is designed to be easily replaced when the need emerges. In order to properly implement the modularity concept, the dynamic library code is typically structured around the interface, the set of functions that exposes the module's functionality to the outer world, whereas the internals of the dynamic library are typically kept away from the prying eyes of the library users.

As the luck would have it, the static libraries are typically designed to provide the "heart and soul" of the dynamic libraries. Regardless of how precious the static library contribution is to the overall functionality of its host dynamic library, the rules of designing the dynamic libraries stipulate that they should export (i.e., make visible) only the bare minimum required for library to communicate with the outer world.

As a direct consequence of such design rules (as you will see in the following chapters), the visibility of the static library symbols ends up being subdued. Instead of remaining globally visible (which they were immediately after the linking completed), the static library symbols immediately become either demoted into the private ones or may even become stripped out (i.e., completely eliminated from the list of dynamic library symbols).

On the other hand, a peculiar yet very important detail is that the dynamic libraries enjoy the complete autonomy over their local symbols. In fact, several dynamic libraries may be loaded into the same process, each dynamic library featuring local symbols that have the same names as the other dynamic libraries' local symbols. Yet the linker manages to avoid any naming conflicts.

The allowed existence of multiple instances of the same-named symbols may lead to a number of unwanted consequences. One scenario is known as the *multiple instances of singleton class paradox*, which will be illustrated in more detail in Chapter 10.

Counterindicated Use Case Scenarios

Say you have a piece of code that provides certain functionality, and you must decide whether or not to encapsulate it in the form of a static library. Here are some typical scenarios in which the static library case is counterindicated:

- When linking the static library requires linking several dynamic libraries (except maybe the libc), then the static library probably should not be used, and the matching dynamic library option should be favored.

    The matching dynamic library option may mean one of the following:

    - The existing dynamic library version of the same library should be used.

    or

    - The library source code (if available) should be rebuilt to create the dynamic library.

    or

    - The available static library should be dismantled into the object files, which (except in a few rare cases) may be used in the build project that builds the dynamic library.

    This is completely analogous to the situation that happens when a person with special needs (special eating habits, or special medical/environmental conditions requirements) decides to stay at friend's house when visiting the town in which the friend lives. In order for the friend to accommodate the guest's special needs, he needs to significantly rearrange his everyday life in order to make unusual extra trips to the specialty food stores, or provide special conditions which he himself does not really need in his everyday life. It makes far more sense for the visitor to take a more independent role, such as getting a hotel room or to arranging the support for his specific needs; and once his own references are resolved, get in touch with the friend whose town he is visiting.

- If the functionality you implement requires the existence of single instance of a class (singleton pattern), following the good dynamic library design practices will ultimately lead to the strong suggestion to encapsulate your code in a dynamic instead of a static library. The rationale behind this was explained in the previous paragraph.

    A good real life example of this scenario is the design of a logging utility. It typically features a single instance of a class visible to a variety of functionality modules, specializing in serializing all possible log statements and sending the log stream to the recording medium (stdout, hard disk or network file, etc.).

    If the functionality modules are implemented as dynamic libraries, hosting the logger class in another dynamic library is strongly suggested.

Specific Rules of Linking Static Libraries

Linking static libraries in Linux adheres to the following set of rules:

- Linking static libraries happens sequentially, one static library by one.

- Linking static libraries starts from the last static library on the list of static libraries passed to the linker (from command line or through the makefile), and goes backwards, toward the first library on the list.

- The linker searches the static libraries in detail, and of all the object files contained in the static library it links in only the object file, which contains symbols that are really needed by the client binary.

As a result of these specific rules, it is *sometimes required to specify the same static library more than once on the same list of static libraries passed to the linker*. The chances of this happening increase when a static library provides several unrelated sets of functionalities.

Converting Static to Dynamic Library

Static library can be converted to dynamic library fairly simply. All you need to do is the following:

- Use the archiver (ar) tool to extract all the object files from the library, like

  ```
  $ ar -x <static library>.a
  ```

  which results with the collection of object files extracted from the static library into the current folder.

  On Windows, you may use the lib.exe tool which is available through the Visual Studio console. Based on the MSDN online documentation (http://support.microsoft.com/kb/31339) it is possible to extract at least one object file (you first need to list the static library contents, which can be also achieved by using the lib.exe tool).

- Build the dynamic library from the set of the extracted object files to the linker.

This recipe works in *almost* all cases. The special cases in which additional requirements must be satisfied are featured next.

Static Libraries Issues on 64-bit Linux

Using the static libraries on 64-bit Linux comes with an interesting corner case scenario. Here is the outline:

- Linking the static library into the executable does not differ from doing the same thing on 32-bit Linux.

- However, linking the static library into the shared library requires that the static library be built with either the -fPIC compiler flag (suggested by the compiler's error printout) or with the -mcmodel=large compiler flag.

This is quite interesting scenario.

First, a mere mentioning of the -fPIC compiler flag in the context of static libraries may be a bit confusing. As I will discuss in the next chapter dealing with dynamic libraries, using the -fPIC flag has been traditionally associated with building dynamic libraries.

It is a popular belief that passing the -fPIC flag to the compiler is one of the two key requirements strictly required by dynamic libraries, but *never ever* required for compiling static library. Any mention of the -fPIC compiler flag in the context of static libraries is a bit shocking.

As matter of fact, this belief is not exactly correct but it's fairly error-safe. The truth is that the use of -fPIC flag is not the decisive factor of whether the static or dynamic library will be created; it is the -*shared* linker flag.

Back to the harsh reality. The true reason why the compiler insists on compiling the static library with the -fPIC flag is that on 64-bit platform the range of address offsets cannot be covered by the usual compiler assembler constructs in which the 32-bit registers are used. The compiler needs a kick of sort (the use of -fPIC or the -mcmodel=large compiler flags) in order to implement the same code with the 64-bit registers.

Resolving the Problem In Real Life Scenarios

It is not completely impossible to get the software package designed much before the era of 64-bit operating systems, in which the static library was built without the -fPIC (or -mcmodel=large) flag. Also, the folks who deliver their static libraries are not necessarily the superstars of dealing with the issues related to compilers/linkers/libraries/ (unlike the guys who complete reading this book ;). If you have had luck (as I have) in getting the static library from the third party developers who were not aware of this particular scenario, there is some bad news: there is no easy workaround for this kind of problem.

Trying to dismantle the static library into the object file does not change the situation even for a tiny bit; the object files haven't been compiled with the compiler flags required for this particular scenario, and no library conversion magic can help avoiding the need to recompile the static library sources.

The only true solution to this kind of a problem is that someone who has the source code (the code distributor or the end user) modifies the build parameters (edit the Makefile) by adding the required flags to the set of compiler flags.

If that is any consolation, imagine that you don't have the library source code at all. Now, that would be scary, huh?

■ ■ ■

## CHAPTER 6

Designing Dynamic Libraries: Basics

Chapter 5 covered the details of the basic ideas behind the static libraries concept, so now it's time to examine the details of dealing with dynamic libraries. This is important because these details affect the everyday work of the programmer/software designer/software architect.

Creating the Dynamic Library

The compilers and linkers typically provide a rich variety of flags which may ultimately provide a lot of flavors to the process of building a dynamic library. For things to be really interesting, even the simplest, widely used recipe that requires one compiler and one linker flag may not be as plain and simple as it initially looks, and the deeper analysis may uncover a really interesting set of facts. Anyways, let's start from the beginning.

Creating the Dynamic Library in Linux

The process of building dynamic libraries has traditionally consisted of the following minimum set of flags:

- -fPIC compiler flag
- -shared linker flag

The following simple example demonstrates the process of creating the dynamic library out of two source files:

```
$ gcc -fPIC -c first.c second.c
$ gcc -shared first.o second.o -o libdynamiclib.so
```

By Linux convention, dynamic libraries start with the prefix lib and have the filename extension .so.

If you follow this recipe, there's little chance you'll go astray. If these flags are passed to the compiler and linker, respectively, whenever you intend to build a dynamic library, the ultimate result will be the correct and usable dynamic library. However, taking this recipe as undisputed and universal truth is not the right thing to do. More precisely, as much as there is really nothing wrong with passing the -shared flag to the linker, the use of the -fPIC compiler flag is a really intriguing topic, which deserves some extra attention.

The rest of this section will be primarily focused on the Linux side (even though some of the concepts live in Windows, too).

About the -fPIC Compiler Flag

The details about using the -fPIC flag can be best illustrated trough the sequence of following questions and answers.

Question 1: What does -fPIC stand for?

The "PIC" in -fPIC is the acronym for ***position-independent code***. Before the concept of position-independent code came to prominence, it was possible to create dynamic libraries that the loader was capable of loading into the process memory space. However, only the process that first loaded the dynamic library could enjoy the benefits of its presence; all other running processes that needed to load the same dynamic library had no choice other than loading another copy of the same dynamic library into memory. The more processes needed to load a particular dynamic library, the more copies in memory had to exist.

The underlying cause of such limitations was a suboptimal loading procedure design. Upon loading the dynamic library into the process, the loader altered the dynamic library's code (.text) segment in a way that made all the dynamic library's symbols meaningful solely within the realm of the process that loaded the library. Even though this approach was suitable for the most basic runtime needs, the ultimate result was that the loaded dynamic library was irreversibly altered so that it would be fairly hard for any other process to reuse the already loaded library. This original loader design approach is known as ***load-time relocation*** and will be discussed in greater detail in subsequent paragraphs.

The PIC concept was clearly a huge step ahead. By redesigning the loading mechanism to avoid tying the loaded library's code (.text) segment to the memory map of the first process that loaded it, the desired extra functionality notch was achieved by providing the way for multiple processes to seamlessly map to its memory map the already loaded dynamic library.

Question 2: Is the use of the -fPIC compiler flag strictly required to build the dynamic library?

The answer is not unique. On 32-bit architecture (X86), it is not required. If not specified, however, the dynamic library will conform to the older load-time relocation loading mechanism in which only the process that loads the dynamic library first will be able to map it into its process memory map.

On 64-bit architectures (X86_64 and I686), the simple omission of the -fPIC compiler flag (in an attempt to implement the load-time relocation mechanism) will result with the linker error. A discussion of why this happens and how to circumvent the problem will be provided later in this book. The remedy for this kind of situation is to pass either the -fPIC flag or -mcmodel=large to the compiler.

Question 3: Is the use of the -fPIC compiler flag strictly confined to the domain of dynamic libraries? Can it be used when building the static library?

It is popular belief that the use of the -fPIC flag is strictly confined to the realm of dynamic libraries. The truth is a bit different.

On 32-bit architecture (X86), it does not really matter if you compile the static library with -fPIC flag or not. It will have a certain impact on the structure of the compiled code; however, it will have negligible impact on the linking and overall runtime behavior of the library.

On 64-bit architecture (X86_64 for sure), things are even more interesting.

- The static library linked into the executable may be compiled with or without the -fPIC compiler flag (i.e., it does not matter whether you specify it or not).

However:

- The static library linked into the dynamic library must be compiled with -fPIC flag !!! (Alternatively, instead of the -fPIC flag, you may specify the -mcmodel=large compiler flag.)

   If the static library was not compiled with either of the two flags, the attempt to link it in into the dynamic library results in the linker error shown in Figure 6-1.

```
/usr/bin/ld: ../staticLib/libstaticlinkingdemo.a(testStaticLinking.o):
relocation R_X86_64_32 against `.rodata' can not be used when making a
shared object; recompile with -fPIC
../staticLib/libstaticlinkingdemo.a: could not read symbols: Bad value
```

Figure 6-1. *Linker error*

An interesting technical discussion related to this problem may be found in the following web article: www.technovelty.org/c/position-independent-code-and-x86-64-libraries.html.

Creating the Dynamic Library in Windows

The process of building a simple dynamic library in Windows requires following a fairly simple recipe. The sequence of screenshots (Figures 6-2 through 6-6) illustrates the process of creating the DLL project. Once the project is created, building the DLL requires nothing more than launching the Build command.

Figure 6-2. *The first step in creating Win32 dynamic library (DLL)*

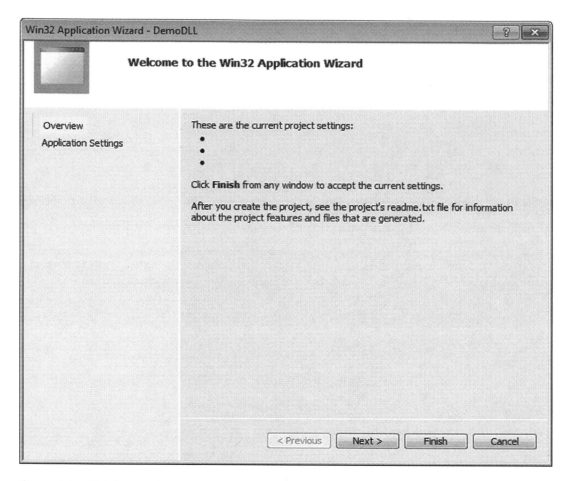

Figure 6-3. *Click the Next button to specify DLL choice*

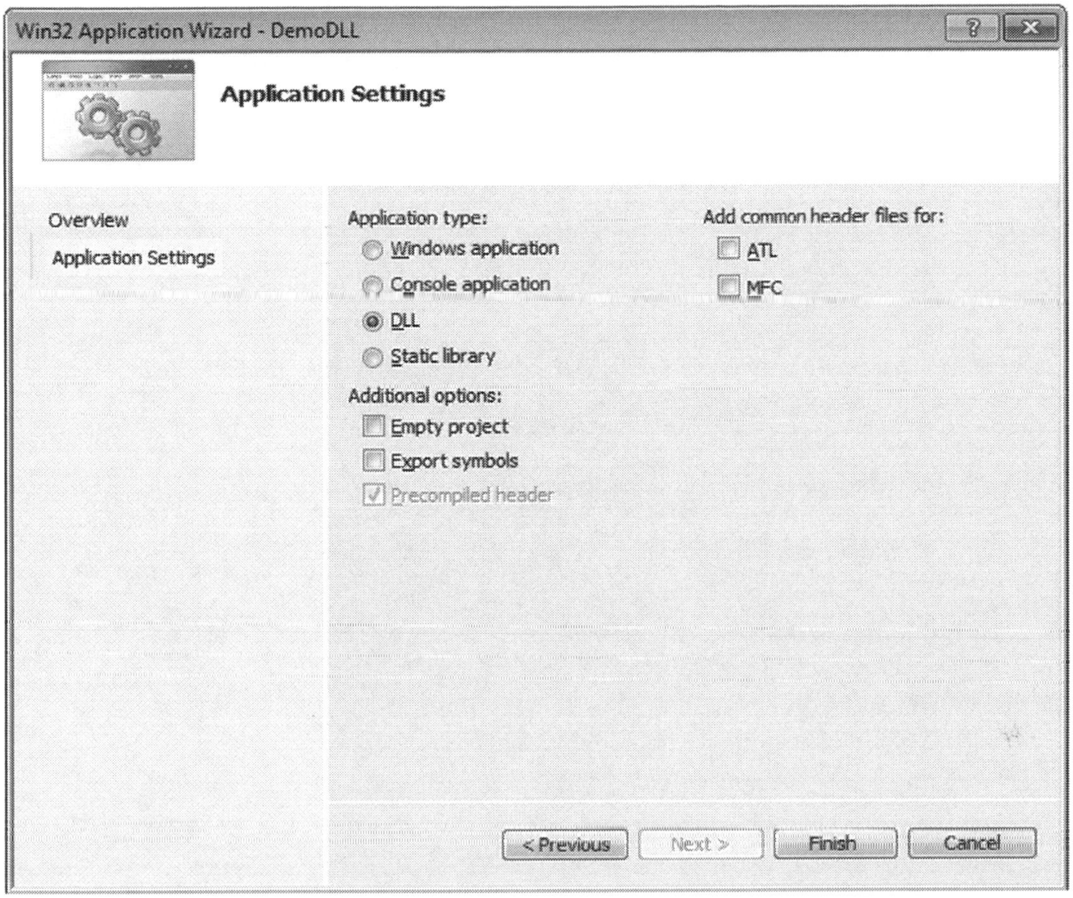

Figure 6-4. *Available Win32 DLL Settings*

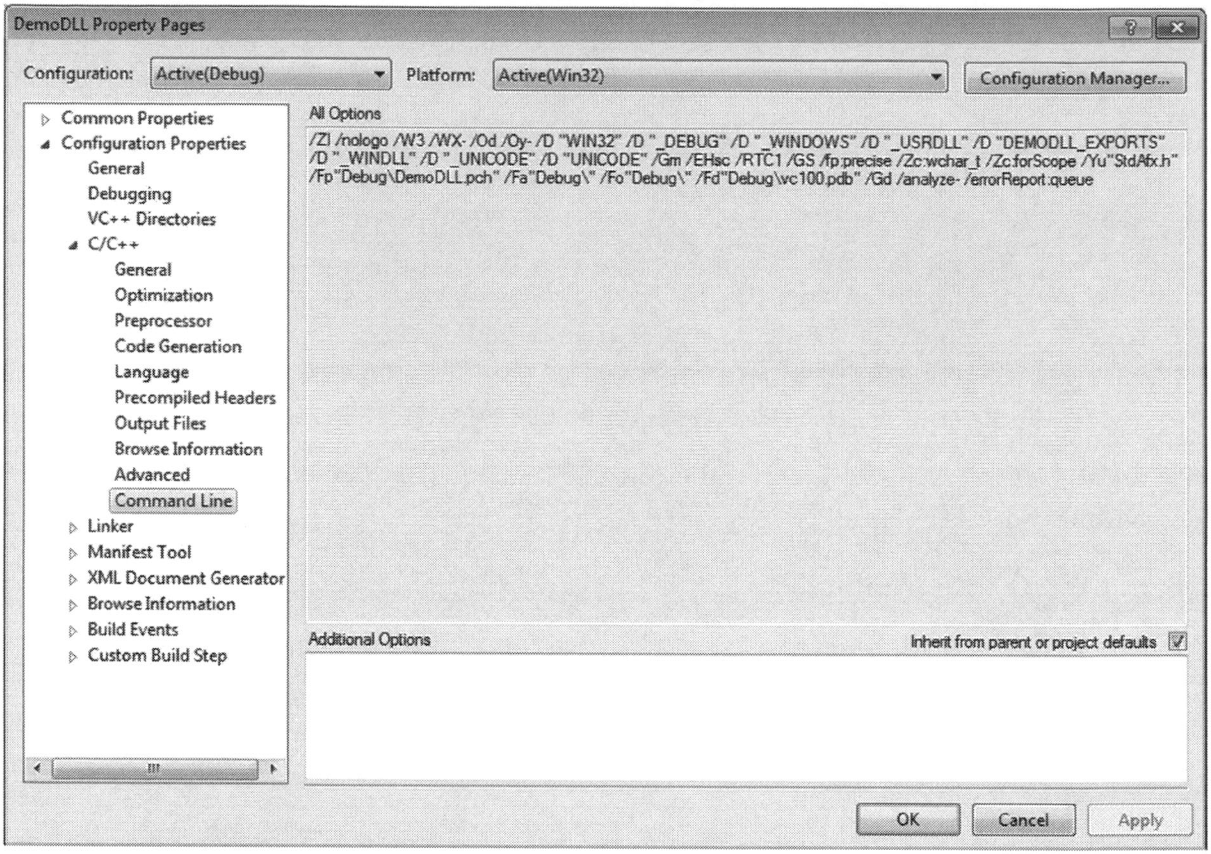

Figure 6-5. *Created DLL compiler flags*

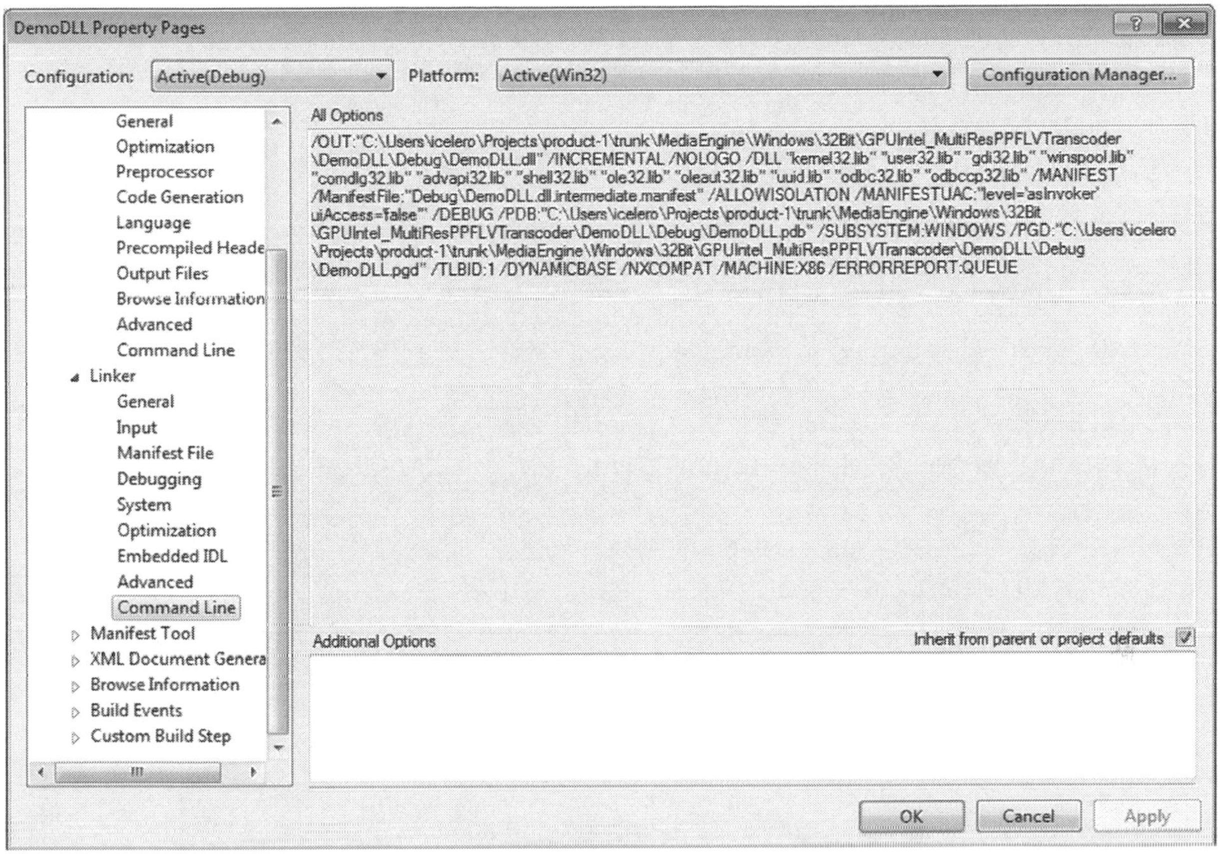

Figure 6-6. *Created DLL linker flags*

Designing Dynamic Libraries

The process of designing a dynamic library in general does not differ much from designing any other piece of software. Given the specific nature of dynamic libraries, there are a few specific points of importance that need to be discussed in detail.

Designing the Binary Interface

By its nature, a dynamic library in general provides a specific functionality to the outer world, the manner of which should minimize the client's involvement in the inner functionality details. The way it is achieved is through the interface, where the client is relieved to the maximum extent of knowing anything it does not need to worry about.

The concept of interfacing, which is omnipresent in the domain of object-oriented programming, obtains an extra flavor in the domain of binary code reuse. As explained in "The Impact of the Binary Reuse Concept" section in Chapter 5, the immutability of the application binary interface (ABI) between the build-time and runtime phases of dynamic linking is the most basic requirement of successful dynamic linking.

At first glance, the design of the ABI does not differ much from the design of the API. The basic meaning of the concept of the interface remains unchanged: a set of functions that need to be made available to the client in order to use the services provided by specialized module.

Indeed, as long as the program is not written in C++, the design effort of the dynamic library's ABI does not require any more thinking than designing the API of a reusable software module. The fact that the ABI is just a set of linker symbols that needs to be loaded at runtime does not make things substantially different.

However, the impact of the C++ language (most notably, the lack of strict standardization) requires additional thinking when designing the dynamic library ABI.

C++ Issues

An unfortunate fact of life is that progress in the domain of programming languages was not symmetrically followed by the design of linkers, or to say precisely, by the strictness of the normative bodies that bring the standards in the software domain. The good reasons for not doing so will be pointed out throughout this section. An excellent article illustrating these issues is "Beginner's Guide To Linkers" at `www.lurklurk.org/linkers/linkers.html`.

Let start with the plain facts and review a few of the issues.

Issue #1: C++ Imposes More Complex Symbol Name Requirements

Unlike the C programming language, the mapping of C++ functions to the linker symbols brings a lot more challenges to the linker design. The object-oriented nature of C++ brings the following extra considerations:

- In general, C++ functions are rarely standalone; they instead tend to be *affiliated* with various code entities.

  The first thing that comes to mind is yes, in C++, functions generally belong to the classes (and as such even have the special name: methods). Additionally, classes (and therefore their methods) may belong to the namespaces. The situation gets even more complicated when templates come to the play.

  In order to uniquely identify the function, the linker must somehow include the function affiliation information to the symbol it creates for the function entry point.

- The C++ overloading mechanism allows that different methods of the same class have the same name, the same return value, but differ in terms of input arguments lists.

  In order to uniquely identify the functions (methods) sharing the same name, the linker must somehow add the information about the input arguments to the symbol it creates for the function entry point.

The linker design efforts to respond to these substantially more complex requirements resulted in the technique known as ***name mangling.*** In a nutshell, name mangling is the process of combining the function name, the function's affiliation information, and the function's list of arguments to create the final symbol name. Usually, the function affiliation is prepended (prefixed), whereas the function signature information is appended (postfixed) to the function name.

The major source of trouble is that the name mangling conventions are not uniquely standardized, and to this day remain vendor-specific. The Wikipedia article (`http://en.wikipedia.org/wiki/Name_mangling#How_different_compilers_mangle_the_same_functions`) illustrates the differences in name mangling implementations across different linkers. As stated in the article, plenty of factors other than just the ABI play a role in implementing the mangling mechanism (exception handling stack, layout of virtual tables, structure and stack frame padding). Given the multitude of various requirements, the Annotated C++ Reference Manual even recommends maintaining individual mangling schemes.

C-STYLE FUNCTIONS

When using the C++ compiler, interesting things happen when using C-style functions. Even though the C functions do not require mangling, the linker by default creates mangled names for them. In cases when it is desired to avoid mangling, a special keyword must be applied in order to suggest to the linker to not apply mangling.

The technique is based on using the `extern "C"` keyword. When a function is declared (typically in a header file) in the following way

```
#ifdef __cplusplus
extern "C"

{
#endif // __cplusplus

int myFunction(int x, int y);

#ifdef __cplusplus

}
#endif // __cplusplus
```

the ultimate result is that linker creates its symbol deprived of any mangling. Later in this chapter the section about exporting ABI will contain a more detailed explanation of why is this technique is a very important one.

Issue #2: Static Initialization Order Fiasco

One of the legacies of the C languages is that linker can handle fairly simply initialized variables, be it simple data types or the structures. All that linker needs to do is to reserve the storage in the `.data` section and write the initial value into that location. In the domain of the C language, the order in which the variables are initialized is generally of no particular importance. All that matters is that the variables initialization be completed before the program starts.

In C++, however, the data type is an object in general, and its initialization is completed at runtime through the process of object construction, which is completed when the class constructor method completes its execution. Obviously, the linker needs to do far many more things in order to initialize the C++ objects. To facilitate the linker's job, the compiler embeds into the object file the list of all constructors that need to be executed for a specific file, and stores this information into the specific object file segment. At link time, the linker examines all the object files and combines these construction lists into the final list that will be executed at runtime.

It is important to mention at this point that the linkers do observe the order of executing the constructors based on the inheritance chain. In other words, it is guaranteed that the base class constructor will be executed first, followed by the constructors of derived classes. This much of a logic embedded into the linker is sufficient for most of the possible scenarios.

The linker, however, is not indefinitely smart. There is unfortunately a whole category of cases in which a programmer does not in any way deviate from the C++ syntax rules, yet the linker's limited logic nevertheless causes very nasty crashes that happen before the program is loaded, way before any debugger can catch it.

The typical scenario of this kind happens when the initialization of an object relies on some other object being initialized beforehand. I will first explain the underlying mechanism of the problem, and then suggest ways for the programmer to avoid them. In circles of C++ programmers this class of problems is typically referred to as a static initialization order fiasco.

■ **Note**   An excellent illustration of the problem and the solution is presented in Scott Meyer's quintessential "Effective C++" book ("Item 47: Ensure that non-local static objects are initialized before they're used").

Problem Description

Non-local static objects are the instances of a C++ class whose visibility scope exceeds the boundaries of a class. More specifically, such objects may be one of the following:

- Defined at global or namespace scope

- Declared static in a class

- Defined static at file scope

Such objects are routinely initialized by the linker before the program starts running. For each of such objects, the linker maintains the list of constructors required to create such object, and executes them in the order specified by the chain of inheritance.

Unfortunately, this is the only object initialization ordering scheme that the linker recognizes and implements. Now is the time for the special twist in the whole story.

Let's assume that one of these objects depends on some other object being initialized beforehand. Assume for example that you have two static objects:

- Object A (instance of class a), which initializes the network infrastructure, queries the list of available networks, initializes the sockets, and establishes the initial connection with the authentication server.

- Object B (instance of class b), which is required to send the message over the network to the remote authentication server, by calling the interface methods on the instance of class b.

Obviously, the correct order of initialization is that object B gets initialized after object A. It is obvious that violating the order of object initialization has a very real potential for wreaking havoc. Even if the designers have been careful enough to envision cases when the initialization is not completed (i.e., checking the pointer values before making the actual calls), the best that can happen is that class B's task does not get completed when expected.

As a matter of fact, there is no rule that dictates the order in which the initialization of static objects will happen. Attempts to implement the algorithm that would examine the code recognize such scenarios and suggest the correct ordering to the linker have proven to belong to the category of problems that are very hard to solve. The presence of other C++ language features (templates) only aggravates the path to the problem solution.

As the ultimate consequence, the linker may decide to initialize the non-local static object in any order. To make things worse, the linker's decision of which order to follow may depend on an unimaginable number of unrelated runtime circumstances.

In real life, such problems are scary for a variety of reasons. First, they are hard to track as they result in crashes happening before the process loading is connected, much before the debugger can be of any help. Additionally, the crash occurrences may not be persistent; the crash may happen every now and then, or in some scenarios every time with different symptoms.

Avoiding the Problem

Even though the problem is not for the faint of the heart, there is a way to avoid the ugly mess. The linker rules do not specify the order of initializing the variables, but the order is very precisely specified for the static variables declared inside a function body. Namely, the object declared as static inside the function (or class method) is initialized when its definition is encountered for the first time during a call to that function.

The solution to this problem becomes obvious. The instances should not be kept free-roaming in the data memory. Instead, they should be

- Declared as static variables inside a function.

- A function should be conveniently used as the only way to access such a variable (returning the reference to the object, for example) defined static at file scope.

In summary, the following two possible solutions are traditionally applied toward solving these kinds of problems:

- **SOLUTION 1:** Provide the custom implementation of the _init() method, a standard method called immediately when the dynamic library is loaded, in which a class static method instantiates the object, thus forcing initialization by the construction. Consequently, the custom implementation of the standard _fini(), a standard method called immediately before the dynamic library is unloaded, may be provided in which the object deallocation may be completed.

- **SOLUTION 2:** Replace direct access to such object with a call to a custom function. Such function will contain a static instance of the C++ class, and will return the reference to it. Before the first access, a variable declared static will be constructed, ensuring that its initialization will happen before the first actual call. The GNU compiler as well as the C++11 standard guarantees that this solution is thread safe.

Issue #3: Templates

The concept of templates is introduced with the purpose of eliminating duplicated and possibly scattered implementations of the same algorithms that mutually differ only in the data type on which the algorithm operates. As useful the concept is, it introduces additional problems to the linking procedure.

The essence of the problem is that different specializations of templates have completely different machine code representations. As luck would have it, once written, the template may be specialized in about gazillion ways, depending on how the template user wanted to use it. The following template

```
template <class T>
T max(T x, T y)
{
  if (x>y) { return x;}
  else     { return y;}
}
```

may be specialized for as many data types that support comparison operator (simple data types ranging from char all the way to double are the immediate candidates).

When the compiler encounters the template, it needs to materialize it into some form of machine code. But, it can't be done until all other source files have been examined to figure out which particular specialization took place in the code. As this may be relatively easy for the case of standalone applications, the task requires some serious thinking when the template is exported by a dynamic library.

There are two general approaches to solving these kinds of problems:

- The compiler can generate all possible template specializations and create weak symbols for each of them. The complete explanation of the weak symbols concept can be found in the discussion about linker symbol types. Note that the linker has the freedom to discard weak symbols once it determines that they are actually not needed in the final build.

- The alternative approach is that the linker does not include the machine code implementations of any of the template specializations up until the very end. Once everything else is completed, the linker may examine the code, determine exactly which specializations are really needed, invoke the C++ compiler to create the required template specializations, and finally, insert the machine code into the executable. This approach is favored by the Solaris C++ compiler suite.

Designing the Application Binary Interface

In order to minimize potential troubles, improve the portability to different platforms, and even enhance the interoperability between the modules created by different compilers, it is highly recommended to practice the following guidelines.

Guideline #1: Implement the Dynamic Library ABI as a Set of C-style Functions

There are a plenty of good reasons of why this advice makes a lot of sense. For example, you can

- Avoid various issues based on C++ vs. linker interaction

- Improve cross-platform portability

- Improve interoperability between the binaries produced by different compilers. (Some of the compilers tend to produce binaries that can be used by the other compilers. Notable examples are the MinGW and Visual Studio compilers.)

In order to export the ABI symbols as C-style functions, use the extern "C" keyword to direct the linker to not apply the name mangling on these symbols.

Guideline #2: Provide the Header File Carrying the Complete ABI Declaration

The "complete ABI declaration" means not only function prototypes, but also the preprocessor definitions, structures layouts, etc.

Guideline #3: Use Widely-Supported Standard C Keywords

More specifically, using your project-specific data type definitions, or platform-specific data types, or anything that is not universally supported across different compilers and/or different platforms is nothing but an invitation for problems down the road. So, try not to act as a fancy-smart-wizz guy; try instead to write your code as plainly and simply as possible.

Guideline #4: Use a Class Factory Mechanism (C++) or Module (C)

If the inner functionality of the dynamic library is implemented by a C++ class, it still does not mean that you should violate guideline #1. Instead, you should follow the so-called class factory approach (Figure 6-7).

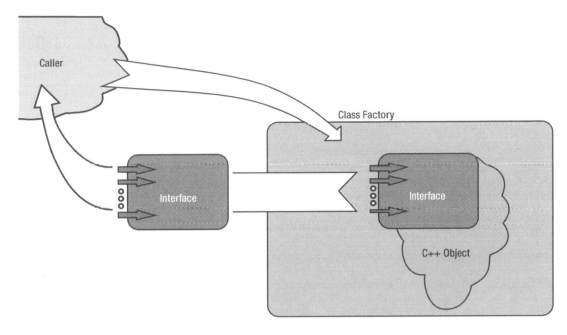

Figure 6-7. *The class factory concept*

The class factory is a C-style function that represents one or more C++ classes to the outer world (similar to a Hollywood agent who represents many star actors in negotiations with film studios).

As a rule, the class factory has intimate knowledge of the C++ class layout, which is typically achieved by declaring it as a static method of the same C++ class.

When called by the interested client, the class factory creates an instance of the C++ class it represents. In order to keep the details of the C++ class layout away from the prying eyes of the client, it never forwards the instance of the class back to the caller. Instead, it casts the C++ class to a C-style interface and casts the pointer to the created C++ object as the interface pointer.

Of course, in order for this scheme to function correctly, the C++ class represented by the class factory is mandated to implement the export interface. In the particular case of C++, it means that the class should publicly inherit the interface. That way, casting the class pointer to the interface pointer comes very naturally.

Finally, this scheme requires that a certain allocation tracking mechanism keeps track of all the instances allocated by class factory function. In Microsoft Component Object Model (COM) technology, the reference counting makes sure that the allocated object gets destroyed when it's no longer being used. In other implementations, it is suggested to keep the list of pointers to the allocated objects. At termination time (delineated by a call to a cleanup function of a kind), each list element would be deleted, and the list finally cleaned up.

The C equivalent of the class factory is typically referred to as module. It is the body of code that provides the functionality to the outer world through a set of carefully designed interface functions.

The modular design is typical for low-level kernel modules and device drivers, but its application is by no means limited to that particular domain. The typical module exports functions such as Open() (or Initialize()), one or more worker functions (Read(), Write(), SetMode(), etc.), and finally, Close() (or Deinitialize()).

Very typical for modules is the use of handle, the module instance identifier of a kind, very frequently implemented as void pointer, a predecessor of this pointer in C++.

The handle typically gets created within the Open() method and is returned to the caller. In calls to the other module interface methods, the handle is the mandatory first function argument.

In the cases where C++ is not an option, designing the C module is completely viable equivalent to the object-oriented concept of a class factory.

Guideline #5: Export Only the Really Important Symbols

By being modular in nature, the dynamic library should be designed so that its functionality is exposed to the outer world through a cleanly defined set of function symbols (the application binary interface, ABI), whereas the symbols of all other functions used only internally should be accessible to the client executables.

There are several benefits to this approach:

- The protection of the proprietary contents is enhanced.

- The library loading time may be tremendously improved as a result of significant reduction in number of exported symbols.

- The chance of conflicted/duplicated symbols between the different dynamic libraries becomes significantly reduced.

The idea is fairly simple: the dynamic library should export only the symbols of the functions and data that are absolutely needed by whoever loads the library, and all the other symbols should be made invisible. The following section will bring more details about controlling the dynamic library symbols visibility.

Guideline #6: Use Namespaces to Avoid Symbol Naming Collision

By encompassing the dynamic library's code into the unique namespace, you eliminate the chances that the different dynamic libraries feature identically named symbols (the function Initialize() is an excellent example of a function that may likely appear in dynamic libraries of completely different scopes of functionality).

Controlling Dynamic Library Symbols' Visibility

From the high-level perspective, the mechanism of exporting/hiding the linker symbols is solved almost identically in both Windows and Linux. The only substantial difference is that by default all Windows DLL linker symbols are hidden, whereas in Linux all the dynamic library linker symbols are by default exported.

In practice, due to a set of features provided by GCC in an attempt to achieve cross-platform uniformity, the mechanisms of symbol exporting look very similar and do pretty much the same thing, in the sense that ultimately only the linker symbols comprising the application binary interface are exported, whereas all remaining symbols are made hidden/invisible.

Exporting the Linux Dynamic Library Symbols

Unlike in Windows, in Linux all of the dynamic library's linker symbols are exported by default, so they are visible by whoever tries to dynamically link the library. Despite the fact that such default makes dealing with the dynamic libraries easy, keeping all the symbols exported/visible is not the recommended practice for many different reasons. Exposing too much to the customers' prying eyes is never a good practice. Also, loading just the required minimum number of symbols vs. loading a gazillion of them may make a noticeable difference in time required to load the library.

It is obvious that some kind of control over which symbols get exported is needed. Furthermore, since such control is already implemented in Windows DLLs, achieving parallelism would tremendously facilitate the portability efforts.

There are several mechanisms for how control over symbol exporting may be achieved at build time. Additionally, the brute-force approach may be applied by running the strip command-line tool over the dynamic library binary. Finally, it is possible to combine several different methods toward the same goal of controlling the visibility of the dynamic library's symbols.

The Symbol Export Control at Build Time

The GCC compiler provides several mechanisms of setting up the visibility of linker symbols:

METHOD 1: (affecting the whole body of code)

```
-fvisibility compiler flag
```

As stated by the GCC man page (http://linux.die.net/man/1/gcc), by passing the
-fvisibility=hidden compiler flag it is possible to make every dynamic library symbols un-exported/invisible
to whoever tries to dynamically link against the dynamic library.

METHOD 2: (affecting individual symbols only)

```
__attribute__ ((visibility("<default | hidden>")))
```

By decorating the function signature with the attribute property, you instruct the linker to either allow (default) or
not allow (hidden) exporting the symbol.

METHOD 3: (affecting individual symbols or a group of symbols)

```
#pragma GCC visibility [push | pop]
```

This option is typically used in the header files. By doing something like this

```
#pragma visibility push(hidden)
void someprivatefunction_1(void);
void someprivatefunction_2(void);
...
void someprivatefunction_N(void);
#pragma visibility pop
```

you are basically making invisible/unexported all the functions declared in between the #pragma statements.

These three methods may be combined in any way the programmer finds it suitable.

The Other Methods

The GNU linker supports a sophisticated method of dealing with the dynamic library versioning, in which a simple
script file is passed to the linker (through the -Wl,--version-script,<script filename> linker flag). As much as the
original purpose of the mechanism is to specify the version information, it also has the power to affect the symbol
visibility. The simplicity with which it accomplishes the task makes this technique the most elegant way of controlling
the symbol visibility. More details about this technique can be found in Chapter 11 in the sections discussing the
Linux libraries versioning control.

The Symbol Export Control Demo Example

In order to illustrate the visibility control mechanism, I've created a demo project in which two otherwise
identical dynamic libraries have been built with different visibility settings. The libraries are appropriately named
libdefaultvisibility.so and libcontrolledvisibility.so. After the libraries are built, their symbols are
examined by using the nm utility (which is covered in detail in Chapters 12 and 13).

The Default Symbols Visibility Case

The source code of libdefaultvisibility.so is shown in Listing 6-1.

Listing 6-1. libdefaultvisibility.so

```
#include "sharedLibExports.h"

void mylocalfunction1(void)
{
        printf("function1\n");
}

void mylocalfunction2(void)
{
        printf("function2\n");
}

void mylocalfunction3(void)
{
        printf("function3\n");
}

void printMessage(void)
{
        printf("Running the function exported from the shared library\n");
}
```

The examination of the symbols present in the built library binary brings no surprises, as the symbols of all functions are exported and visible, as shown in Figure 6-8.

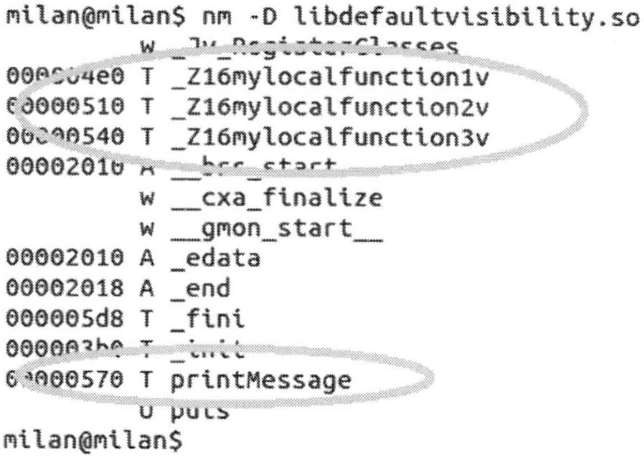

```
milan@milan$ nm -D libdefaultvisibility.so
            w __Jv_RegisterClasses
00000004e0 T _Z16mylocalfunction1v
00000510 T _Z16mylocalfunction2v
00000540 T _Z16mylocalfunction3v
00002010 A __bss_start
            w __cxa_finalize
            w __gmon_start__
00002010 A _edata
00002018 A _end
000005d8 T _fini
000003b0 T _init
00000570 T printMessage
            U puts
milan@milan$
```

Figure 6-8. *All library symbols are originally exported/visible*

The Controlled Symbols Visibility Case

In the case of a dynamic library in which you want to control the symbol visibility/exportability, the -fvisibility compiler flag was specified in the project Makefile, as shown in Listing 6-2.

Listing 6-2. The -fvisibility Compiler Flag

```
...

#
# Compiler
#
INCLUDES        = $(COMMON_INCLUDES)
DEBUG_CFLAGS    = -Wall -g -O0
RELEASE_CFLAGS  = -Wall -O2
VISIBILITY_FLAGS = -fvisibility=hidden -fvisibility-inlines-hidden

ifeq ($(DEBUG), 1)
CFLAGS          = $(DEBUG_CFLAGS) -fPIC $(INCLUDES)
else
CFLAGS          = $(RELEASE_CFLAGS) -fPIC $(INCLUDES)
endif

CFLAGS          += $(VISIBILITY_FLAGS)

COMPILE         = g++ $(CFLAGS)

...
```

When the library is built solely with this particular symbol visibility setting, the examination of the symbols indicated that the function symbols haven't been exported (Figure 6-9).

```
milan@milan$ nm -D libcontrolledvisibility.so
         w _Jv_RegisterClasses
00002010 A __bss_start
         w __cxa_finalize
         w __gmon_start__
00002010 A _edata
00002018 A _end
00000538 T _fini
00000304 T _init
         U puts
milan@milan$
```

Figure 6-9. All library symbols are now hidden

Next, when the function signature decoration with the visibility attributes is applied, as shown in Listing 6-3, the net effect is that the function declared with the __attribute__ ((visibility("default"))) becomes visible (Figure 6-10).

Listing 6-3. The Function Signature Decoration with the Visibility Attributes Applied

```
#include "sharedLibExports.h"

#if 1
#define FOR_EXPORT __attribute__ ((visibility("default")))
#else
#define FOR_EXPORT
#endif

void mylocalfunction1(void)
{
        printf("function1\n");
}

...etc...

//
// also supported:
//              FOR_EXPORT void printMessage(void)
// but this is not supported:
//      void printMessage FOR_EXPORT (void)
// nor this:
//              void printMessage(void) FOR_EXPORT
//
// i.e. attribute may be declared anywhere
// before the function name

void FOR_EXPORT printMessage(void)
{
        printf("Running the function exported from the shared library\n");
}


milan@milan$ nm -D libcontrolledvisibility.so
         w _Jv_RegisterClasses
00002010 A __bss_start
         w __cxa_finalize
         w __gmon_start__
00002010 A _edata
00002018 A _end
00000538 T _fini
0000022. T _tnit
000004ce T printMessage
         U puts
milan@milan$
```

Figure 6-10. *Visibility control applied to function printMessage*

Using the strip Utility

Another mechanism of controlling the symbols visibility is available. It is not as sophisticated and it is not programmable. Instead, it is implemented by running a `strip` command-line utility (Figure 6-11). This approach is far more brutal, as it has the power to completely erase any information about any of the library symbols, to the point that none of the usual symbol examination utilities will be able to see any of the symbols at all, be it in the `.dynamic` section or not.

```
milan@milan$ strip --strip-symbol _Z16mylocalfunction1v libcontrolledvisibility.so
milan@milan$ strip --strip-symbol _Z16mylocalfunction2v libcontrolledvisibility.so
milan@milan$ strip --strip-symbol _Z16mylocalfunction3v libcontrolledvisibility.so
milan@milan$ nm libcontrolledvisibility.so
00001f28 a _DYNAMIC
00001ff4 a _GLOBAL_OFFSET_TABLE_
         w _Jv_RegisterClasses
00001f18 d __CTOR_END__
00001f14 d __CTOR_LIST__
00001f20 d __DTOR_END__
00001f1c d __DTOR_LIST__
000006bc r __FRAME_END__
00001f24 d __JCR_END__
00001f24 d __JCR_LIST__
00002010 A __bss_start
         w __cxa_finalize@@GLIBC_2.1.3
00000520 t __do_global_ctors_aux
000003a0 t __do_global_dtors_aux
0000200c d __dso_handle
         w __gmon_start__
00000457 t __i686.get_pc_thunk.bx
00002010 A _edata
00002018 A _end
00000558 T _fini
0000032c T _init
00002010 b completed.6159
00002014 b dtor_idx.6161
00000420 t frame_dummy
000004f0 T printMessage
         U puts@@GLIBC_2.0
milan@milan$
```

Figure 6-11. *Using the strip utility to eliminate certain symbols*

■ **Note**   More information about the `strip` utility can be found in Chapter 13.

Exporting the Windows Dynamic Library Symbols

In Linux, all the linker symbols found in the dynamic library are by default accessible by the client executables. In Windows, however, this is not the case. Instead, only the symbols that have been properly exported become visible to the client executable. The important part of enforcing this limitation is the use of a separate binary (import library) during the build-time phase, which contains only the symbols planned to be exported.

The mechanism of exporting the DLL symbols is fortunately completely under the programmer's control. In fact, there are two supported mechanisms of how the DLL symbols may be declared for export.

Using the __declspec(dllexport) Keyword

This mechanism is standardly provided by Visual Studio. Check the "Export symbols" checkbox in the new project creation dialog, as shown in Figure 6-12.

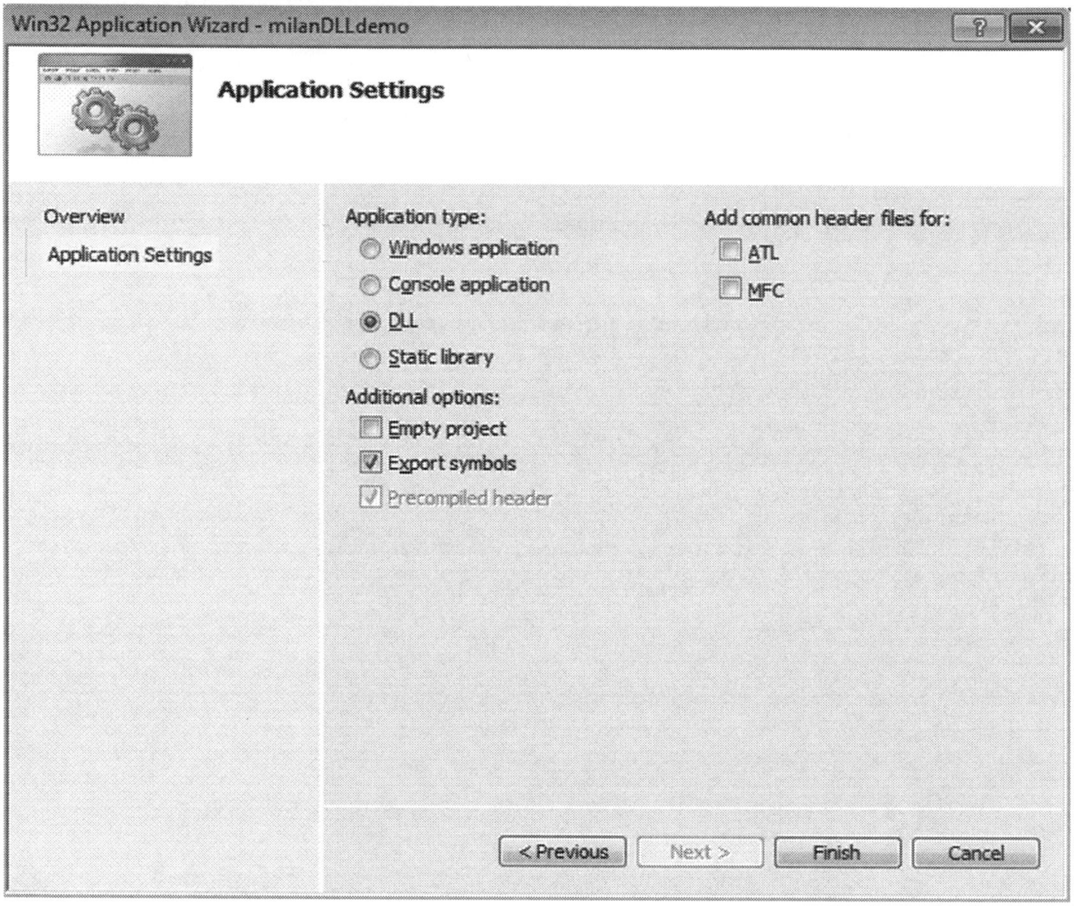

Figure 6-12. *Selecting the "Export symbols" option in Win32 DLL Wizzard dialog*

Here you specify that you want the project wizard to generate the library export header containing the snippets of code looking somewhat like Figure 6-13.

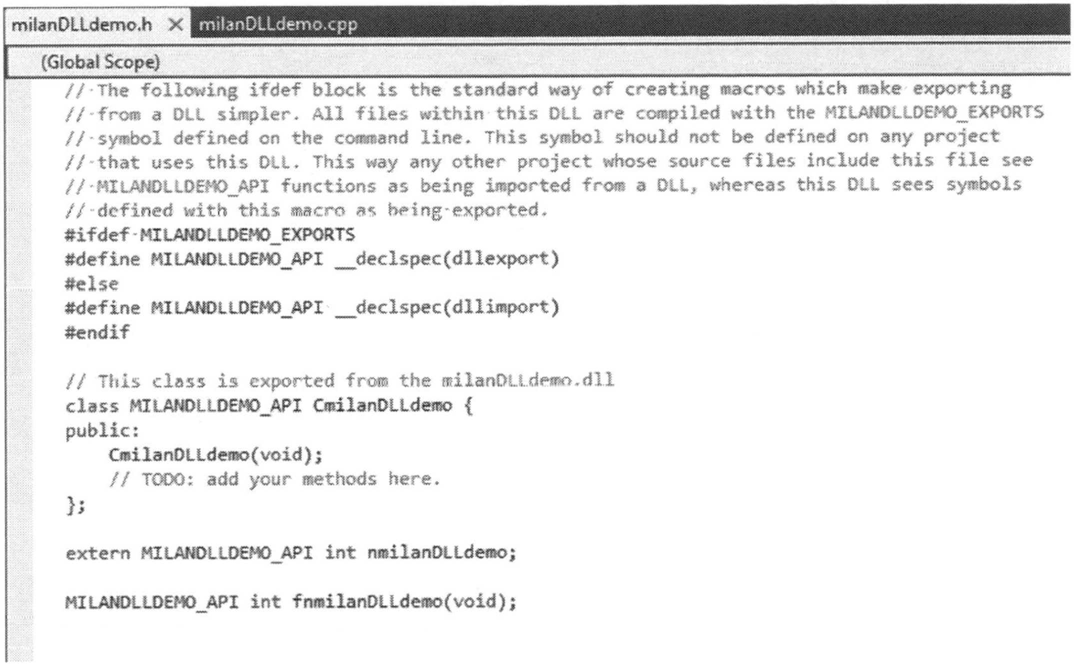

```
milanDLLdemo.h  ×   milanDLLdemo.cpp
(Global Scope)
    // The following ifdef block is the standard way of creating macros which make exporting
    // from a DLL simpler. All files within this DLL are compiled with the MILANDLLDEMO_EXPORTS
    // symbol defined on the command line. This symbol should not be defined on any project
    // that uses this DLL. This way any other project whose source files include this file see
    // MILANDLLDEMO_API functions as being imported from a DLL, whereas this DLL sees symbols
    // defined with this macro as being exported.
    #ifdef MILANDLLDEMO_EXPORTS
    #define MILANDLLDEMO_API __declspec(dllexport)
    #else
    #define MILANDLLDEMO_API __declspec(dllimport)
    #endif

    // This class is exported from the milanDLLdemo.dll
    class MILANDLLDEMO_API CmilanDLLdemo {
    public:
        CmilanDLLdemo(void);
        // TODO: add your methods here.
    };

    extern MILANDLLDEMO_API int nmilanDLLdemo;

    MILANDLLDEMO_API int fnmilanDLLdemo(void);
```

Figure 6-13. *Visual Studio generates project-specific declaration of __declspec(dllexport) keywords*

As Figure 6-13 shows, the export header can be used both inside the DLL project and by the client executable project. When used within the DLL project, the project-specific macro evaluates to the __declspec(dllexport) keyword within the DLL project, whereas within the client executable project it evaluates to __declspec(dllimport). This is enforced by Visual Studio, which automatically inserts the preprocessor definition into the DLL project (Figure 6-14).

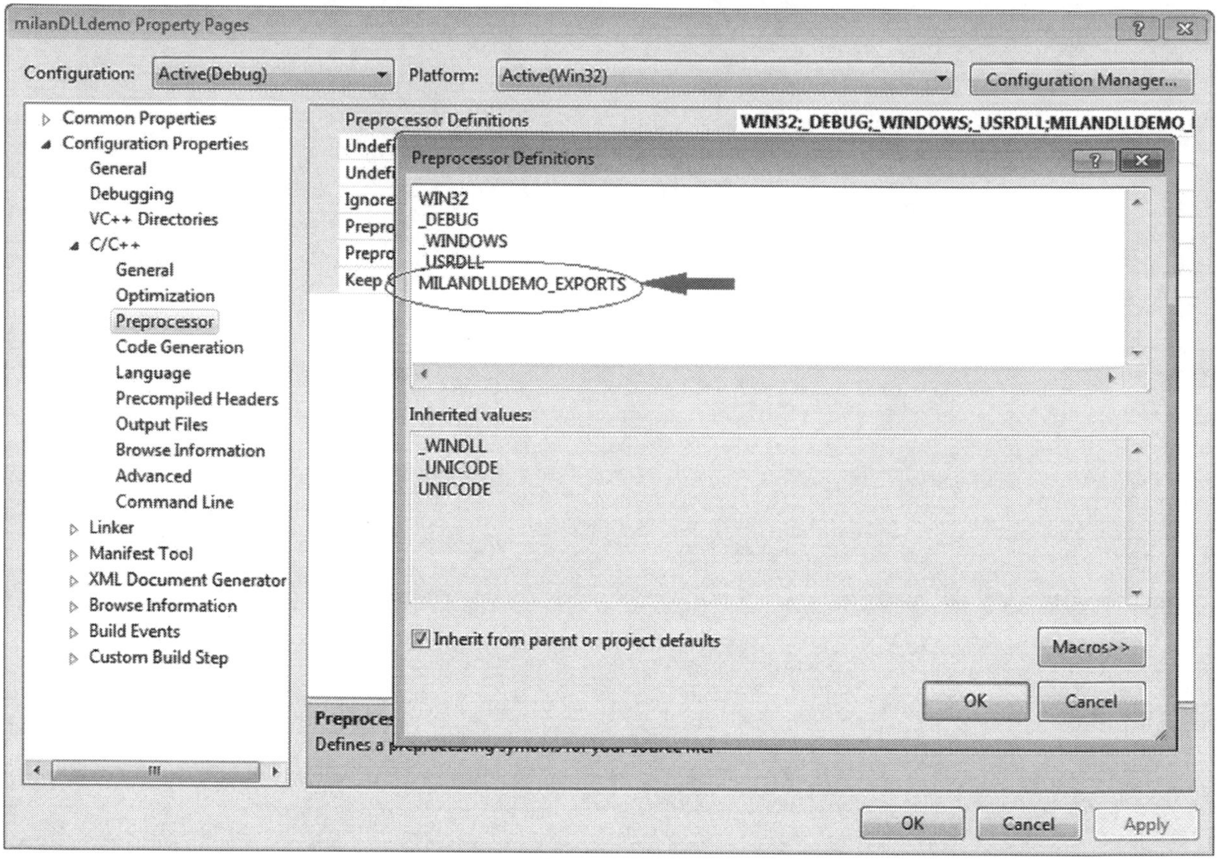

Figure 6-14. *Visual Studio automatically generates the project-specific preprocessor definition*

When the project-specific keyword that evaluates to __declspec(dllexport) is added to the function declaration, the function linker symbol becomes exported. Otherwise, omitting such a project-specific keyword is guaranteed to prevent the exporting of the function symbol. Figure 6-15 features two functions, of which only one is declared for export.

```
#include "stdafx.h"
#include "milanDLLdemo.h"


// This is an example of an exported variable
/* MILANDLLDEMO_API */ int nmilanDLLdemo=0;

// This is an example of an exported function.
MILANDLLDEMO_API int fnmilanDLLdemo(void)
{
    return 256;
}

int notExportingThisFunction(void)
{
    return -1;
}
```

Figure 6-15. *Visual Studio automatically generates example of using project-specific symbol export control keyword*

Now is the perfect moment to introduce the Visual Studio dumpbin utility which you may use to analyze the DLL in the search for exported symbols. It is the part of Visual Studio tools, and can be used only by running the special Visual Studio Tools command prompt (Figure 6-16).

Figure 6-16. *Launching Visual Studio command prompt to access the collection of binary analysis command-line tools*

Figure 6-17 shows what the dumpbin tool (invoked with the /EXPORT flag) reports about the symbols exported by your DLL.

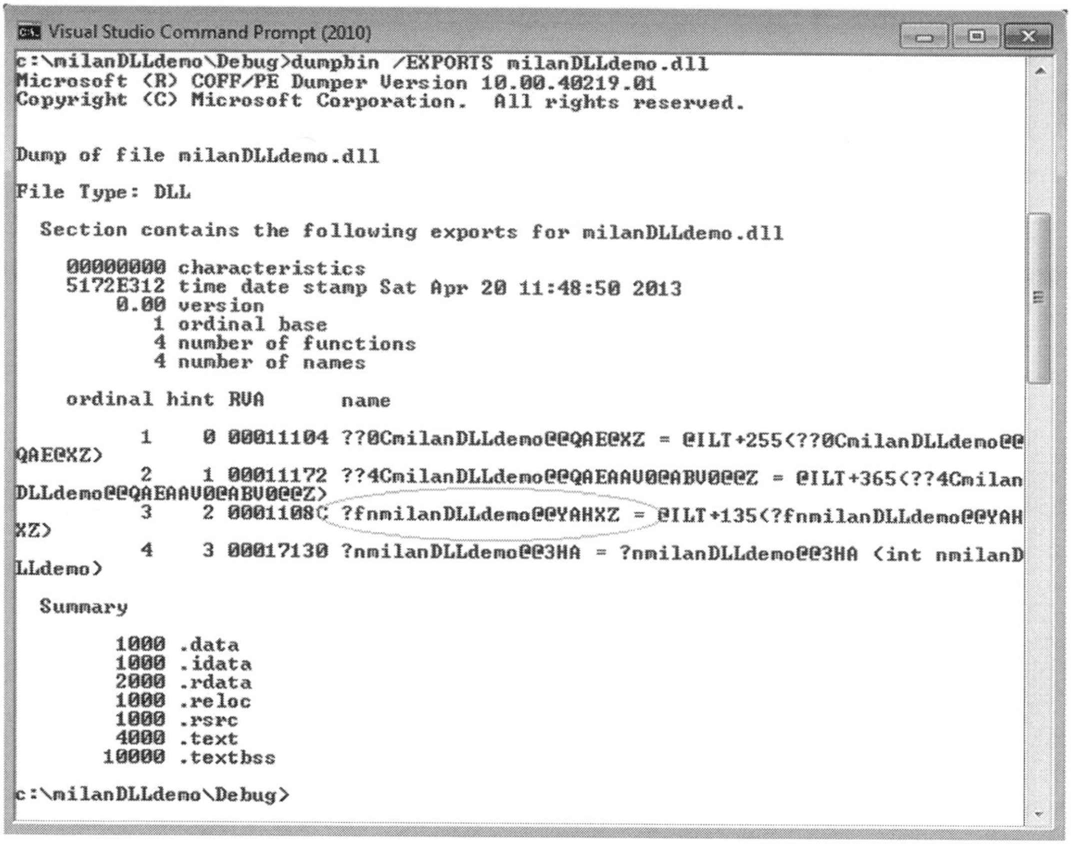

Figure 6-17. *Using dumpbin.exe to view the list of DLL exported symbols*

Obviously, the symbol of function declared with the project-specific export symbol ends up being exported by the DLL. However, the linker processed it according to the C++ guidelines, which use name mangling. The client executables usually do not have problems interpreting such symbols, but if they do, you may declare the function as extern "C", which will result with the function symbol following the C-style convention (Figure 6-18).

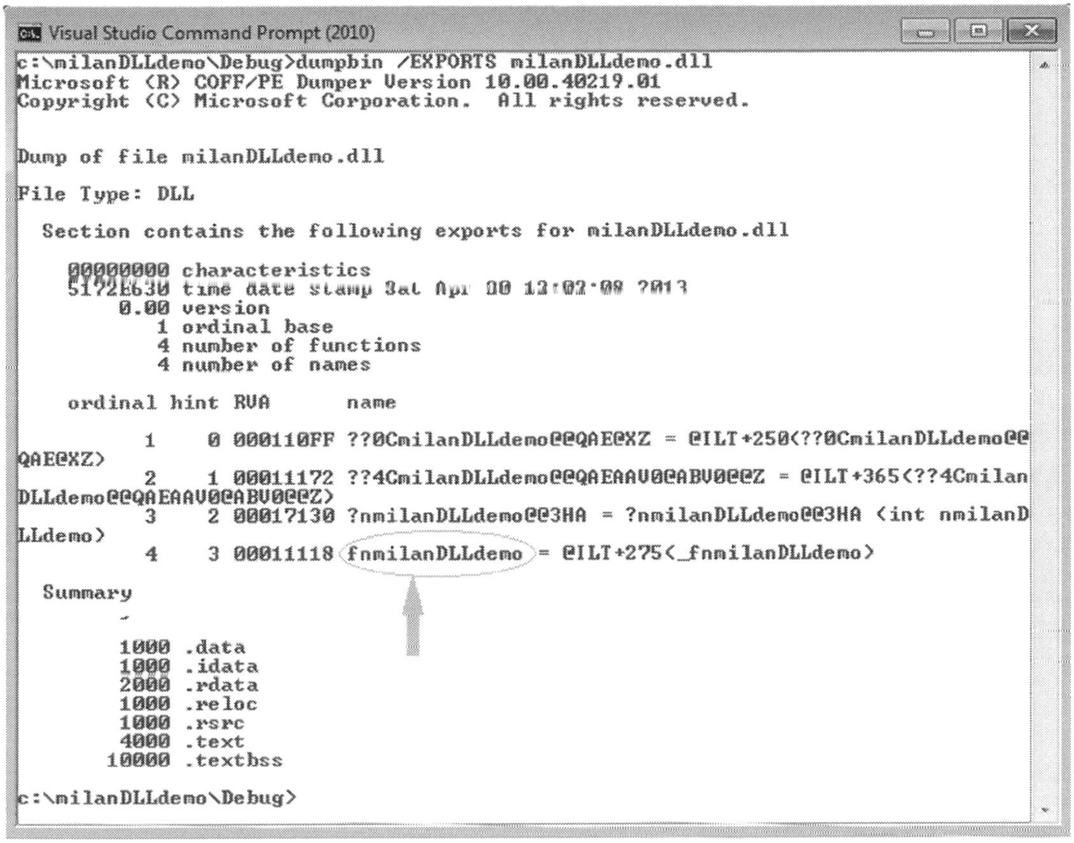

```
Visual Studio Command Prompt (2010)                                    ─ □ ✕

c:\milanDLLdemo\Debug>dumpbin /EXPORTS milanDLLdemo.dll
Microsoft (R) COFF/PE Dumper Version 10.00.40219.01
Copyright (C) Microsoft Corporation.  All rights reserved.


Dump of file milanDLLdemo.dll

File Type: DLL

  Section contains the following exports for milanDLLdemo.dll

    00000000 characteristics
    5172E630 time date stamp Sat Apr 00 13:02:09 2013
        0.00 version
           1 ordinal base
           4 number of functions
           4 number of names

    ordinal hint RVA      name

           1    0 000110FF ??0CmilanDLLdemo@@QAE@XZ = @ILT+250(??0CmilanDLLdemo@@
QAE@XZ)
           2    1 00011172 ??4CmilanDLLdemo@@QAEAAV0@ABV0@@Z = @ILT+365(??4Cmilan
DLLdemo@@QAEAAV0@ABV0@@Z)
           3    2 00017130 ?nmilanDLLdemo@@3HA = ?nmilanDLLdemo@@3HA (int nmilanD
LLdemo)
           4    3 00011118 fnmilanDLLdemo = @ILT+275(_fnmilanDLLdemo)

  Summary

        1000 .data
        1000 .idata
        2000 .rdata
        1000 .reloc
        1000 .rsrc
        4000 .text
       10000 .textbss

c:\milanDLLdemo\Debug>
```

Figure 6-18. *Declaring the function as* extern "C"

Using the Module-definition File (.def)

The alternative way of controlling the export of DLL symbols is through the use of module-definition (.def) files. Unlike the previously described mechanism (based on the __declspec(dllexport) keywords), which can be specified through the project creation wizard by checking the "Export symbols" checkbox, the use of a module-definition file requires some more explicit measures.

For starters, if you plan on using the .def file, it is recommended to not check the "Export symbols" checkbox. Instead, use the File ➡ New menu to create a new definition (.def) file. If this step is completed correctly, the project settings will indicate that the module definition file is officially part of the project, as shown in Figure 6-19.

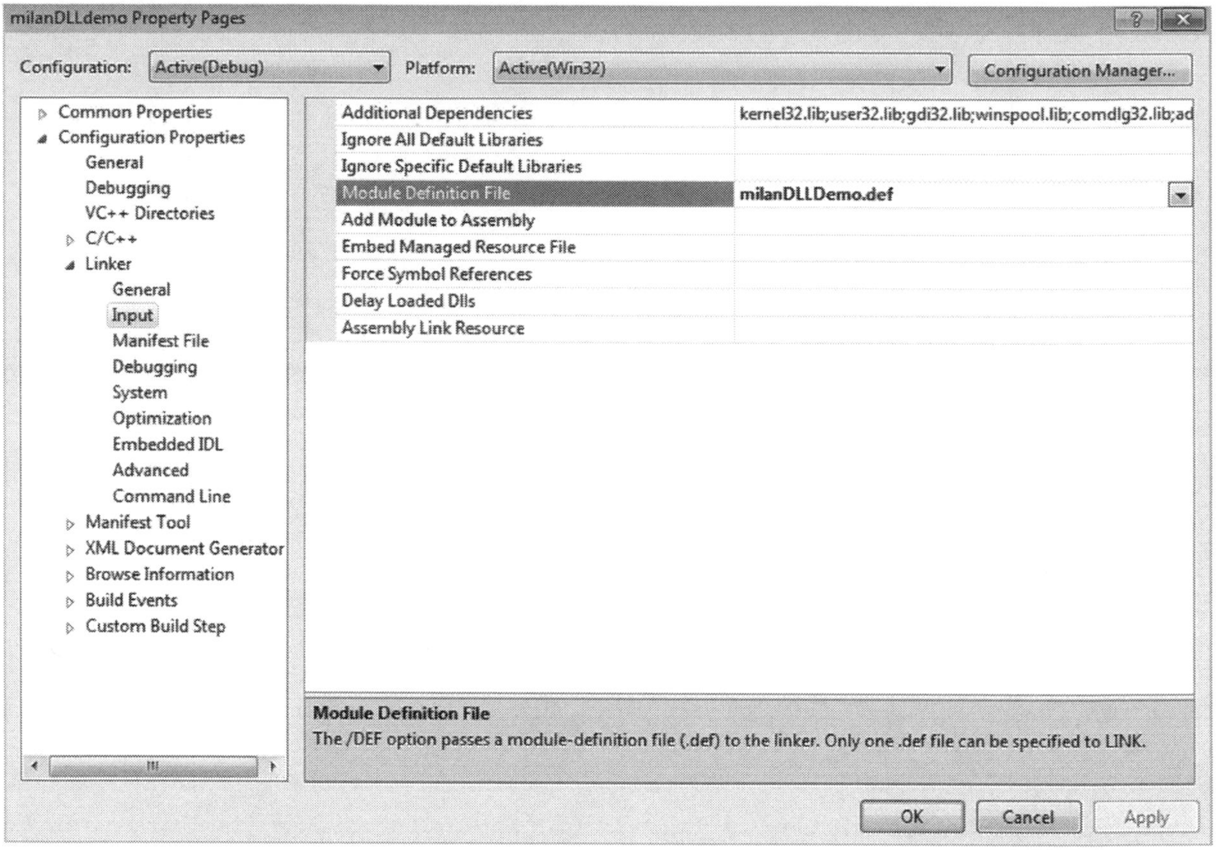

Figure 6-19. *Module-definition (.def) file is officially part of the project*

Alternatively, you may manually write the `.def` file, add it manually to the list of project source files, and finally, manually edit the linker properties page to look as shown in Figure 6-19. The module-definition file that specifies the demo function for export looks like Figure 6-20.

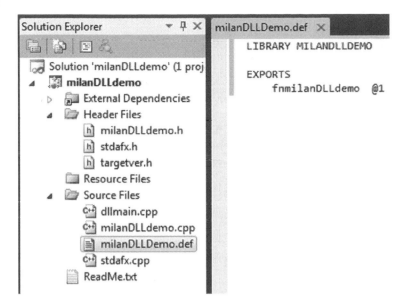

Figure 6-20. *Module-definition file example*

Under the EXPORTS line it may contain as many lines as there are functions whose symbols you plan on exporting.

An interesting detail is that the use of the module-definition file results in function symbols exported as C-style functions, without you needing to declare the function as extern "C". Whether this is an advantage or disadvantage depends on the personal preferences and design circumstances.

One particular advantage of using the module definition (.def) files as the method of exporting the DLL symbols is that in certain cross-compiling scenarios the non-Microsoft compilers tend to support this option.

One such example is using the MinGW compiler, which compiles an open source project (e.g. ffmpeg) to create Windows DLLs and associated .def files. In order for DLL to be dynamically linked at build time, you need to use its import library, which was unfortunately not generated by the MinGW compiler.

Fortunately, the Visual Studio Tools provide the lib.exe command-line utility which can generate the import library file based on the contents of the .def file (Figure 6-20). The lib tool is available through the Visual Studio Tools command prompt. The example in Figure 6-21 illustrates how the tool was used after the cross-compiling session in which the MinGW compiler run on Linux produced the Windows binaries (but did not supply the import libraries).

```
X:\MilanFFMpegWin32Build>dir *.def
 Volume in drive X is VBOX_VBoxShared
 Volume Serial Number is 9AE7-0879

 Directory of X:\WinFFMpegBuiltOnLinux

02/14/2013  11:51 AM             7,012 avcodec-53.def
02/14/2013  11:51 AM               115 avdevice-53.def
02/14/2013  11:51 AM             5,107 avfilter-2.def
02/14/2013  11:51 AM             5,119 avformat-53.def
02/14/2013  11:51 AM             4,762 avutil-51.def
02/14/2013  11:51 AM               232 postproc-51.def
02/14/2013  11:51 AM               155 swresample-0.def
02/14/2013  11:51 AM             7,084 swscale-2.def
               8 File(s)         29,586 bytes
               0 Dir(s)  465,080,082,432 bytes free

X:\MilanFFMpegWin32Build>lib /machine:X86 /def:avcodec-53.def /out:avcodec.lib

Microsoft (R) Library Manager Version 10.00.40219.01
Copyright (C) Microsoft Corporation.  All rights reserved.

   Creating library avcodec.lib and object avcodec.exp

X:\MilanFFMpegWin32Build>lib /machine:X86 /def:avdevice-53.def /out:avdevice.lib

Microsoft (R) Library Manager Version 10.00.40219.01
Copyright (C) Microsoft Corporation.  All rights reserved.

   Creating library avdevice.lib and object avdevice.exp

X:\MilanFFMpegWin32Build>lib /machine:X86 /def:avfilter-2.def /out:avfilter.lib

Microsoft (R) Library Manager Version 10.00.40219.01
Copyright (C) Microsoft Corporation.  All rights reserved.

   Creating library avfilter.lib and object avfilter.exp

X:\MilanFFMpegWin32Build>lib /machine:X86 /def:avformat-53.def /out:avformat.lib

Microsoft (R) Library Manager Version 10.00.40219.01
Copyright (C) Microsoft Corporation.  All rights reserved.

   Creating library avformat.lib and object avformat.exp

X:\MilanFFMpegWin32Build>lib /machine:X86 /def:avutil-51.def /out:avutil.lib
Microsoft (R) Library Manager Version 10.00.40219.01
Copyright (C) Microsoft Corporation.  All rights reserved.

   Creating library avutil.lib and object avutil.exp

X:\MilanFFMpegWin32Build>lib /machine:X86 /def:postproc-51.def /out:postproc.lib

Microsoft (R) Library Manager Version 10.00.40219.01
Copyright (C) Microsoft Corporation.  All rights reserved.

   Creating library postproc.lib and object postproc.exp

X:\MilanFFMpegWin32Build>lib /machine:X86 /def:swresample-0.def /out:swresample.
lib
Microsoft (R) Library Manager Version 10.00.40219.01
Copyright (C) Microsoft Corporation.  All rights reserved.

   Creating library swresample.lib and object swresample.exp

X:\MilanFFMpegWin32Build>lib /machine:X86 /def:swscale-2.def /out:swscale.lib
Microsoft (R) Library Manager Version 10.00.40219.01
Copyright (C) Microsoft Corporation.  All rights reserved.

   Creating library swscale.lib and object swscale.exp

X:\MilanFFMpegWin32Build>
```

Figure 6-21. *Generating import library files for DLLs generated by MingW compiler based on specified module definition (.def) files*

Shortcomings of dealing with the module definition file (.def)

While experimenting with the .def files, the following shortcomings have been perceived:

- *An inability to discern the C++ class methods from C functions:* If within a DLL you have a class, and the class has a method of the same name as the C function you specified for export in the .def file, the compiler will report a conflict while trying to figure out which of the two should be exported.

- extern "C" *quirks:* In general, the function declared for export inside the .def file does not need to be declared as extern "C" as the linker will take care that its symbol follows the C convention. However, if you decide to nevertheless decorate the function as extern "C", make sure to do it in both header file and in the source .cpp file (the latter of which should not be normally required). Failing to do so will somehow confuse the linker, and the client application will not be able to link your exported function symbol. For the problem to be harder, the dumpbin utility output will not indicate any difference whatsoever, making the problem harder to resolve.

Linking Completion Requirements

The dynamic library creation process is a complete build procedure, as it involves both the compiling and the linking stage. In general, the linking stage gets completed once every linker symbol has been resolved, and this criterion should be observed regardless of whether the target is an executable or dynamic library.

In Windows, this rule is strictly enforced. The linking process is never considered complete and the output binary never gets created until every dynamic library symbol had been resolved. The complete list of dependent libraries is searched up until the last symbol reference has been resolved.

In Linux, however, this rule is a bit twisted by default when building the dynamic libraries, as it is allowed that the linking of dynamic library gets completed (and the binary file created) even though not all of the symbols have been resolved.

The reason behind allowing this deviation from the strictness of the otherwise solid rule is that it is implicitly assumed that the symbols missing during the linking stage will eventually somehow appear in the process memory map, very likely as a result of some other dynamic library being loaded at runtime. The needed symbols not supplied by the dynamic library are marked as undefined ("U").

As a rule, if the expected symbol for whatever reason does not appear in the process' memory map, the operating system tends to neatly report the cause by printing the text message on the stderr stream, specifying the missing symbol.

This flexibility in the Linux rules of linking the dynamic libraries has been proven on a number of occasions as a positive factor, allowing certain very complex linking limitations to be effectively overcome.

--no-undefined Linker Flag

Despite the fact that linking the dynamic libraries is far more relaxed in Linux by default, the GCC linker supports the option of establishing the linking strictness criteria matching those followed by the Windows linker.

Passing the --no-undefined flag to the gcc linker will result with the unsuccessful build if each and every symbol is not resolved at build time. This way, the Linux default of tolerating the presence of unresolved symbols is effectively reverted into the Windows-like strict criteria.

Note that when invoking the linker through the gcc, the linker flags must be preceded by the -Wl, prefix, such as:

```
$ gcc -fPIC <source files> -l <libraries> -Wl,--no-undefined -o <shlib output filename>
```

Dynamic Linking Modes

The decision to link the dynamic library may be made at different stages of the program life cycle. In some scenarios, you know up front that your client binary will need to load certain dynamic library no matter what. In other scenarios, the decision about loading certain dynamic library comes as the result of runtime circumstances, or user preferences set at runtime. Based on when the decision about dynamically linking is actually made, the following dynamic linking modes can be distinguished.

Statically Aware (Load-Time) Dynamic Linking

In all the discussions so far, I have implicitly assumed this particular scenario. Indeed, it happens very frequently that the need for a particular dynamic library's functionality is needed from the very moment when the program is started all the way to the program's termination, and this fact is known up front. In this scenario, the build procedure requires the following items.

At compile time:

- The dynamic library's export header file, specifying everything that pertains to the library's ABI interface

At link time:

- The list of dynamic libraries required by the project

- The paths to the dynamic library binaries needed by the client binary in order to set up the list of expected library symbols.

  For more details about how the paths may be specified, please check the "Build-Time Library Location Rules" section.

- Optional linker flags specifying the details of the linking process

Runtime Dynamic Linking

The whole beauty of the dynamic linking feature is the ability for the programmer to determine at runtime whether the need for a certain dynamic library really exists and/or which particular library needs to be loaded.

Many times, the design requires that a number of dynamic libraries exist, each of which supports the identical ABI, and that only one of them gets loaded depending on the user's choice. A typical example for such a scenario is the multiple language support where, based on the user's preferences, the application loads the dynamic library that contains all the resources (strings, menu items, help files) written in the user's language of choice.

In this scenario, the build procedure requires the following items.

At compile time:

- The dynamic library's export header file, specifying everything that pertains to the library's ABI interface

At link time:

- At least the filename of the dynamic library to be loaded. The exact path of the dynamic library filename typically gets resolved implicitly, by relying on the set of priority rules governing the choice of paths in which the library binary is expected to be found at runtime.

All major operating systems provide a simple set of API functions that allow programmer to fully exploit this precious feature (Table 6-1).

Table 6-1. *API Functions*

Purpose	Linux Version	Windows Version
Library Loading	dlopen()	LoadLibrary()
Finding Symbol	dlsym()	GetProcAddress()
Library Unloading	dlclose()	FreeLibrary()
Error Reporting	dlerror()	GetLastError()

Regardless of the operating system and/or programming environment, the typical paradigm of using these functions can be described by the following pseudocode sequence:

```
1) handle = do_load_library("<library path>", optional_flags);
   if(NULL == handle)
      report_error();

2) pFunction = (function_type)do_find_library_symbol(handle);
   if(NULL == pFunction)
   {
      report_error();
      unload_library();
      handle = NULL;
      return;
   }

3) pFunction(function arguments); // execute the function

4) do_unload_library(handle);
   handle = NULL;
```

Listings 6-4 and 6-5 provide simple illustrations of runtime dynamic loading.

Listing 6-4. Linux Runtime Dynamic Loading

```
#include <stdlib.h>
#include <stdio.h>
#include <dlfcn.h>

#define PI (3.1415926536)

typedef double (*PSINE_FUNC)(double x);

int main(int argc, char **argv)
{
    void *pHandle;

    pHandle = dlopen ("libm.so", RTLD_LAZY);
    if(NULL == pHandle) {
        fprintf(stderr, "%s\n", dlerror());
        return -1;
    }
```

```
    PSINE_FUNC pSineFunc = (PSINE_FUNC)dlsym(pHandle, "sin");
    if (NULL == pSineFunc) {
       fprintf(stderr, "%s\n", dlerror());
       dlclose(pHandle);
       pHandle = NULL;
       return -1;
    }

    printf("sin(PI/2) = %f\n", pSineFunc(PI/2));
    dlclose(pHandle);
    pHandle = NULL;
    return 0;
}
```

Listing 6-5 illustrates the Windows runtime dynamic loading, in which we try to load the DLL, locate the symbols of functions *DllRegisterServer()* and/or *DllUnregisterServer()* and execute them.

Listing 6-5. Windows Runtime Dynamic Loading

```
#include <stdio.h>
#include <Windows.h>

#ifdef __cplusplus
extern "C"
{
#endif // __cplusplus
typedef HRESULT (*PDLL_REGISTER_SERVER)(void);
typedef HRESULT (*PDLL_UNREGISTER_SERVER)(void);
#ifdef __cplusplus
}
#endif // __cplusplus

enum
{
  CMD_LINE_ARG_INDEX_EXECUTABLE_NAME = 0,
  CMD_LINE_ARG_INDEX_INPUT_DLL,
  CMD_LINE_ARG_INDEX_REGISTER_OR_UNREGISTER,
  NUMBER_OF_SUPPORTED_CMD_LINE_ARGUMENTS
} CMD_LINE_ARG_INDEX;

int main(int argc, char* argv[])
{
  HINSTANCE dllHandle = ::LoadLibraryA(argv[CMD_LINE_ARG_INDEX_INPUT_DLL]);
  if(NULL == dllHandle)
  {
    printf("Failed loading %s\n", argv[CMD_LINE_ARG_INDEX_INPUT_DLL]);
    return -1;
  }
```

```
  if(NUMBER_OF_SUPPORTED_CMD_LINE_ARGUMENTS > argc)
  {
    PDLL_REGISTER_SERVER pDllRegisterServer =
        (PDLL_REGISTER_SERVER)GetProcAddress(dllHandle, "DllRegisterServer");
    if(NULL == pDllRegisterServer)
    {
      printf("Failed finding the symbol \"DllRegisterServer\"");
      ::FreeLibrary(dllHandle);
      dllHandle = NULL;
      return -1;
    }
    pDllRegisterServer();
  }
  else
  {
    PDLL_UNREGISTER_SERVER pDllUnregisterServer =
        (PDLL_UNREGISTER_SERVER)GetProcAddress(dllHandle, "DllUnregisterServer");
    if(NULL == pDllUnregisterServer)
    {
      printf("Failed finding the symbol \"DllUnregisterServer\"");
      ::FreeLibrary(dllHandle);
      dllHandle = NULL;
      return -1;
    }
    pDllUnregisterServer();
  }

  ::FreeLibrary(dllHandle);
  dllHandle = NULL;
  return 0;
}
```

Dynamic Linking Modes Comparison

There are very few substantial differences between the two modes of dynamic linking. Even though the moment in which the dynamic linking happens differs, the actual mechanism of dynamic linking is completely identical in both cases.

Furthermore, the dynamic library that can be loaded statically aware can also be dynamically loaded at runtime. There are no elements of the dynamic library design that would qualify the library strictly for use in one vs. another scenario.

The only substantial difference is that in the statically aware scenario has one extra requirement that needs to be satisfied: you need to provide the build-time library location. As will be shown in the next chapter, this task requires some finesses that a good software developer needs to be aware of in both Linux and Windows environments.

CHAPTER 7

■ ■ ■

Locating the Libraries

The idea of binary code sharing is at the core of the concept of libraries. Somewhat less obvious is that it typically means that the single copy of the library binary file will reside in a fixed location on a given machine, whereas plenty of different client binaries will need to *locate* the needed library (either at build time or at runtime). In order to address the issue of locating the libraries, a variety of conventions have been devised and implemented. In this chapter, I will discuss the details of these conventions and guidelines.

Typical Library Use Case Scenarios

The use of libraries has proven to be a very powerful way of sharing code across the software community. It is a very common practice that companies with accumulated expertise in certain domains deliver their intellectual property in the form of libraries, which third parties may integrate into their products and deliver to customers.

The practice of using the libraries happens through two distinct use case scenarios. The first use case scenario happens when the developers try to integrate the third-party libraries (static or dynamic) in their product. Another scenario happens when libraries (in this case specifically dynamic libraries) need to be located at runtime in order for the application installed on a client's machine to run properly.

Both use case scenarios introduce the problem of locating the library binary files. The way these problems have been structurally resolved will be described in this chapter.

Development Use Case Scenario

Typically, the third-party package containing the library, export headers, and possibly few extras (such as documentation, online help, package icons, utility applications, code and media samples, etc.) is installed on a predetermined path on the developer's machine. Immediately after, the developer may create a plethora of projects residing at many different paths on her machine.

Obviously, each and every project that requires linking with third-party libraries needs to have access to the library binaries. Otherwise, it would be impossible to finish building the project.

Copying the third-party library to each and every project that a developer may create is definitely a possibility, albeit a very poor choice. Obviously, having copies of the library in the folder of each project that may need it defeats the original idea of code reuse standing behind the concept of libraries.

The acceptable alternative would be to have just one copy of the library binaries, and a set of rules helping the client binary projects to locate it. Such set of rules, commonly referred to as *build time library location rules*, is typically supported by the development platform's linker. These rules basically stipulate how the information about the path of the libraries needed to complete the client binary linking may be passed to the linker.

The build time library location rules are fairly elaborate and tend to come in a variety of options. Each of the major development platforms usually provides very sophisticated set of options of how these rules may be enforced.

It is very important to understand that the *build time library location rules are pertinent to both static and dynamic libraries*. Regardless of the actual differences between linking static vs. dynamic libraries, the linker nevertheless must know the location of the needed library binaries.

End User Runtime Use Case Scenario

Once the developers have integrated the third-party libraries, their products are ready to be delivered to the end clients. Based on the vast variety of design criteria and real-life considerations, the structure of the delivered product may come in the wide variety of choices:

- In the simplest possible case, the product package contains only a single application file. The intended use is that the client simply runs the application.

  This case is fairly trivial. In order to access and run the application, all the user needs to do is to add its path to the overall PATH environment variable. Anybody other than the completely computer illiterate person is capable of completing this simple task.

- In the more complex scenarios, the product package contains a mix of dynamic libraries and one or more utility applications. The dynamic libraries may be either directly forwarded third-party libraries, or may be created by the package vendor, or may be a combination of both.

  The intended use is that a variety of applications dynamically link with the supplied dynamic libraries. The typical examples of such a scenario in the domain of multimedia are the multimedia frameworks such as DirectX or GStreamer, as each of them supplies (or counts on being available at runtime) an elaborate set of dynamic libraries, each providing a certain well-defined set of functionalities.

Much like in the development use case scenario, the meaningful approach to the problem assumes that there will be only one copy of the required dynamic libraries, residing at the path where the installation procedure deployed them. On the other hand, these libraries may be need by a multitude of client binaries (other dynamic libraries or the applications) residing at a plethora of different paths.

In order to structure the process of finding the dynamic libraries binaries at runtime (or slightly before, at load time), a set of *runtime library location rules* needs to be established. The *runtime library location rules* are usually fairly elaborate. Each of the development platforms provides its own flavor of sophisticated options of how these rules may be enforced.

Finally—at the risk of reiterating the obvious—the *runtime library location rules pertain only to the dynamic libraries*. The integration of static libraries is always completed much before the runtime (i.e., at the linking stage of the client binary build process), and there is never a need to locate the static libraries at runtime.

Build Time Library Location Rules

In this section, I will discuss the techniques of providing the build time paths to the library binary files. Beyond the simplest possible move of just furnishing the full path to the linker, there are some extra levels of finesse that deserve your attention.

Linux Build Time Library Location Rules

The important part of the recipe of how the build time library location rules are implemented on Linux belongs to the Linux libraries naming conventions.

Linux Static Library Naming Conventions

The Linux static library filenames are standardly created according to the following pattern:

```
static library filename = lib + <library name> + .a
```

The middle part of the library's filename is the library's actual name, which is used to submit the library to the linker.

Linux Dynamic Library Naming Conventions

Linux features a very elaborate dynamic libraries naming convention scheme. Even though the original intention was to address the library versioning issues, the naming convention scheme impacts the library location mechanisms. The following paragraphs will illustrate the important points.

Dynamic Library Filename vs. Library Name

The Linux dynamic library filenames are standardly created according to the following pattern:

```
dynamic library filename = lib + <library name> + .so + <library version information>
```

The middle part of the library's filename is the library's actual name, which is used to submit the library to the linker, and later on to the build time library search as well as the runtime library search procedures.

Dynamic Library Version Information

The library version information carried by the last part of the library's filename adheres to the following convention:

```
dynamic library version information = <M>.<m>.<p>
```

where each of the mnemonics may represent one or more digits indicating

- *M*: major version
- *m*: minor version
- *p*: patch (minor code change) version

The importance of dynamic library versioning information will be discussed in detail in Chapter 11.

Dynamic Library Soname

By definition, the dynamic library's **soname** can be specified as

```
library soname = lib + <library name> + .so + <library major version digit(s)>
```

For example, the soname of the library *libz.so.1.2.3.4* would be ***libz.so.1***.

The fact that only the major version digits play a role in the library's soname implies that the libraries whose minor versions differ will still be described by the same *soname* value. Exactly how this feature is used will be discussed in the Chapter 11 section dedicated to the topic of dynamic library versions handling.

The library *soname* is typically embedded by the linker into the dedicated ELF field of the library's binary file. The string specifying the library soname is typically passed to the linker through the dedicated linker flag, like so:

```
$ gcc -shared <list of object files> -Wl,-soname,libfoo.so.1 -o libfoo.so.1.0.0
```

The utility programs for examining the contents of the binary files typically provide the option for retrieving the *soname* value (Figure 7-1).

```
milan@milan:~$ readelf -d /lib/i386-linux-gnu/libz.so.1.2.3.4

Dynamic section at offset 0x13ee8 contains 23 entries:
  Tag        Type                         Name/Value
 0x00000001 (NEEDED)                     Shared library: [libc.so.6]
 0x0000000e (SONAME)                     Library soname: [libz.so.1]
 0x0000000c (INIT)                       0x1400
 0x0000000d (FINI)                       0xe668
 0x6ffffef5 (GNU_HASH)                   0x138
 0x00000005 (STRTAB)                     0xa4c
```

Figure 7-1. *Library soname embedded in the library binary's ELF header*

Linker's vs. Human's Perception of Library Name

Please notice that the library name as described by these conventions is not necessarily used in human conversation to denote the library. For example, the library providing the compression functionality on a given machine may reside in the filename *libz.so.1.2.3.4*. According to the library naming convention, this library's name is simply "*z*", which will be used in all the dealings with the linker and loader. From the human communication prospective, the library may be referred to as "*libz*", as for example in the following bug description in a bug tracking system: "Issue 3142: problem with missing *libz* binary". To avoid the confusion, sometimes the library name is also referred to as the *library's linker name*.

Linux Build Time Library Location Rules Details

The build time library path specification is implemented on Linux in the form of so-called -L -l option. The truly correct way of using these two options may be described by the following set of guidelines:

- Break the complete library path into two parts: the folder path and the library filename.

- Pass the folder path to the linker by appending it after the -L linker flag.

- Pass the *library name (linker name)* only to linker by appending it after the -l flag.

For example, the command line for creating the application demo by compiling the file main.cpp and linking in with the dynamic library libworkingdemo.so located in the folder ../sharedLib may look something like this:

```
$ gcc main.o -L../sharedLib -lworkingdemo -o demo
               ^                ^
               |                |
        library folder path   library name only
                              (not the full library filename !)
```

In the cases when the gcc line combines compiling with linking, these linker flags should be prepended with the -Wl, flag, like so:

```
$ gcc -Wall -fPIC main.cpp -Wl,-L../sharedLib -Wl,-lworkingdemo -o demo
```

Beginners' Mistakes: What Can Possibly Go Wrong and How to Avoid It

The typical problems happen to the impatient and inexperienced programmer in *scenarios dealing with dynamic libraries*, when either of the following situations happen:

- The full path to a dynamic library is passed to the -l option (-L part being not used).

- The part of the path is passed through the -L option, and the rest of the path including the filename passed through the -l option.

The linker usually formally accepts these variations of specifying the build time library paths. In the case when the path to static libraries is provided, these kinds of "creative freedoms" do not cause problems down the road.

However, when the path to the dynamic library is passed, the problems introduced by deviating from the truly correct way of passing the library path start showing up at runtime. For example, let's say that a client application *demo* depends on the library libmilan.so, which resides on the developer's machine in the following folder:

/home/milan/mylibs/case_a/libmilan.so.

The client application is successfully built by the following linker command line:

```
$ gcc main.o -l/home/milan/mylibs/case_a/libmilan.so -o demo
```

and runs just fine on the same machine.

Let's assume now that the project is deployed to a different machine and given to a user whose name is "john." When that user tries to run the application, nothing will happen. Careful investigation (techniques of which will be discussed in the Chapters 13 and 14) will reveal that the application requires at runtime the dynamic library libmilan.so (which is OK) but it expects to find it at the path /home/milan/mylibs/case_a/.

Unfortunately, this folder does not exist on the user "john"'s machine!

Specifying relative paths instead of absolute paths may only partially alleviate the problem. If, for example, the library path is specified as relative to the current folder (i.e., ../mylibs/case_a/libmilan.so), then the application on john's machine would run only if the client binary and the required dynamic library are deployed to john's machine in the folder structure that maintains the exact relative positions between the executable and dynamic library. But, if john dares to copy the application to a different folder and tries to execute it from there, the original problem will reappear.

Not only that, but the application may stop working even on the developer's machine where it used to work perfectly. If you decide to copy the application binary at a different path on the developer's machine, the loader will start searching for the library on the paths relative to the point where the app binary resides. Very likely such path will not exist (unless you bother to re-create it)!

The key to understanding the underlying cause of the problem is to know that *the linker and the loader do not equally value the library paths passed under the -L vs. under the -l option.*

In fact, the linker gives far more significance to what you passed under the -l option. More specifically, the part of the path passed through the -L option finds its use only during the linking stage, but plays no role thereafter.

The part specified under the -l option, however, gets imprinted into the library binary, and continues playing important role during the runtime. In fact, when trying to find the libraries required at runtime, the loader first reads in the client binary file trying to locate this particular information.

If you dare to deviate from the strict rule, and pass anything more than just the library filename through the -l option, the application built on milan's machine when deployed and run on john's machine will look for the dynamic library at hardcoded path, which most likely exists only on developer's (milan) machine but not at user's (john) machine. An illustration of this concept is provided in Figure 7-2.

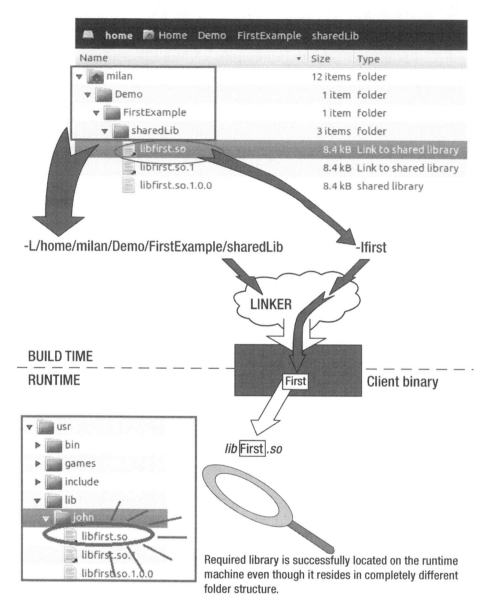

Figure 7-2. *The -L convention plays a role only during library building. The impact of -l convention, however, remains important at runtime, too*

Windows Build Time Library Location Rules

There are several ways how the information about the dynamic library required at link time may be passed to the project. Regardless of the way chosen to specify the build time location rules, the mechanism works for both static and dynamic libraries.

Project Linker Settings

The standard option is to supply the information about the DLL required by the linker as follows:

- Specify the DLL's import library (.lib) file in the list of linker inputs (Figure 7-3).

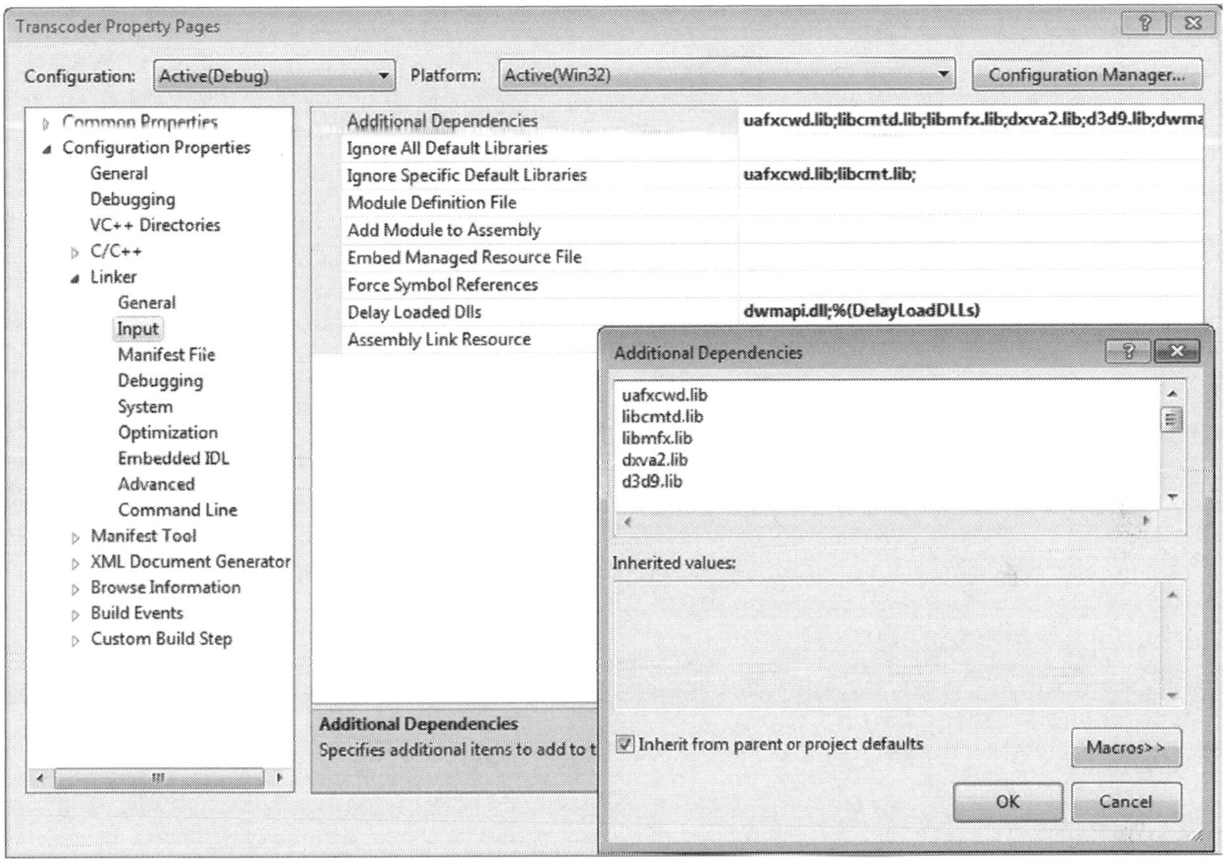

Figure 7-3. *Specify needed libraries in the list of dependencies*

- Add the import library's path to the set of library path directories (Figure 7-4).

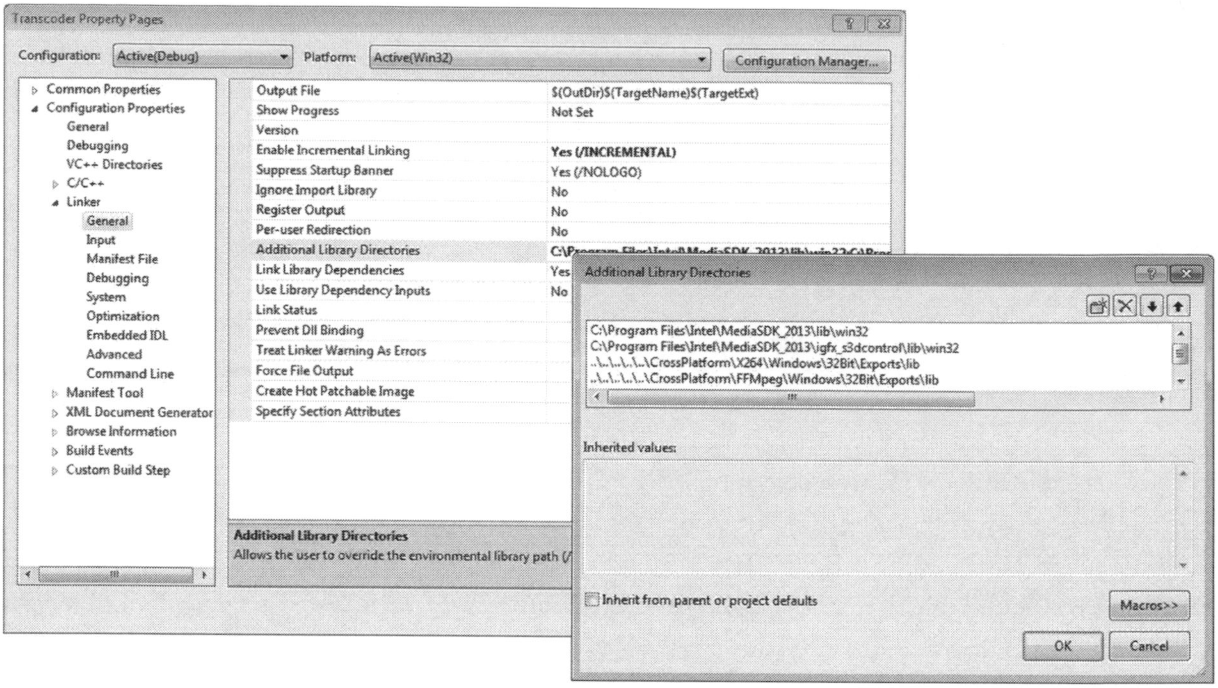

Figure 7-4. Specify the library paths

#pragma Comment

The library requirement may be specified by adding a line like this one to a source file:

```
#pragma comment(lib, "<import library name, full path, or relative path>");
```

Upon encountering this directive, the compiler will insert a library search record in the object file, which will eventually be picked up by the linker. If only the library filename is provided in the double quotes, the search for the library will follow the Windows library search rules. Typically, this option is used to exercise more precision during the process of searching for libraries, and it is hence more usual to specify the library full path and version than otherwise.

One huge advantage of specifying the build-time library requirements in this way is that by being in the source code, it gives the designer the ability to define the linking requirements depending on preprocessor directives. For example,

```
#ifdef CUSTOMER_XYZ
#pragma comment(lib, "<customerXYZ-specific library>");
#else
#ifdef CUSTOMER_ABC
#pragma comment(lib, "<customerABC-specific library>");
#else
#ifdef CUSTOMER_MPQ
#pragma comment(lib, "<customerMPQ-specific library>");
#endif // CUSTOMER_MPQ
#endif // CUSTOMER_ABC
#endif // CUSTOMER_XYZ
```

Implicit Referencing of the Library Project

This option may be used only in special case when both the dynamic library project and its client executable project are the parts of the same Visual Studio solution. If the DLL project is added to the client app project's list of references, the Visual Studio environment will provide everything (automatically and mostly invisibly to the programmer) needed for building and running the app.

For starters, it will pass the complete path of the DLL to the application's linker command line. Finally, it will perform the copying of the DLL to the application's runtime folder (typically Debug for the debug version, and Release for the release version), thus satisfying in the easiest possible way the rules of runtime library location.

Figures 7-5 through 7-8 illustrate the recipe of how to do it, using the example of the solution (SystemExamination) comprised of the two related projects: a SystemExaminer DLL which is statically aware linked in by the SystemExaminerDemoApp application.

I will not specify the DLL dependency by relying on the first method described previously (i.e. by specifying the DLL's import library (.lib) file in the list of linker inputs). This seemingly peculiar and a bit counter-intuitive detail is illustrated by the Figure 7-5.

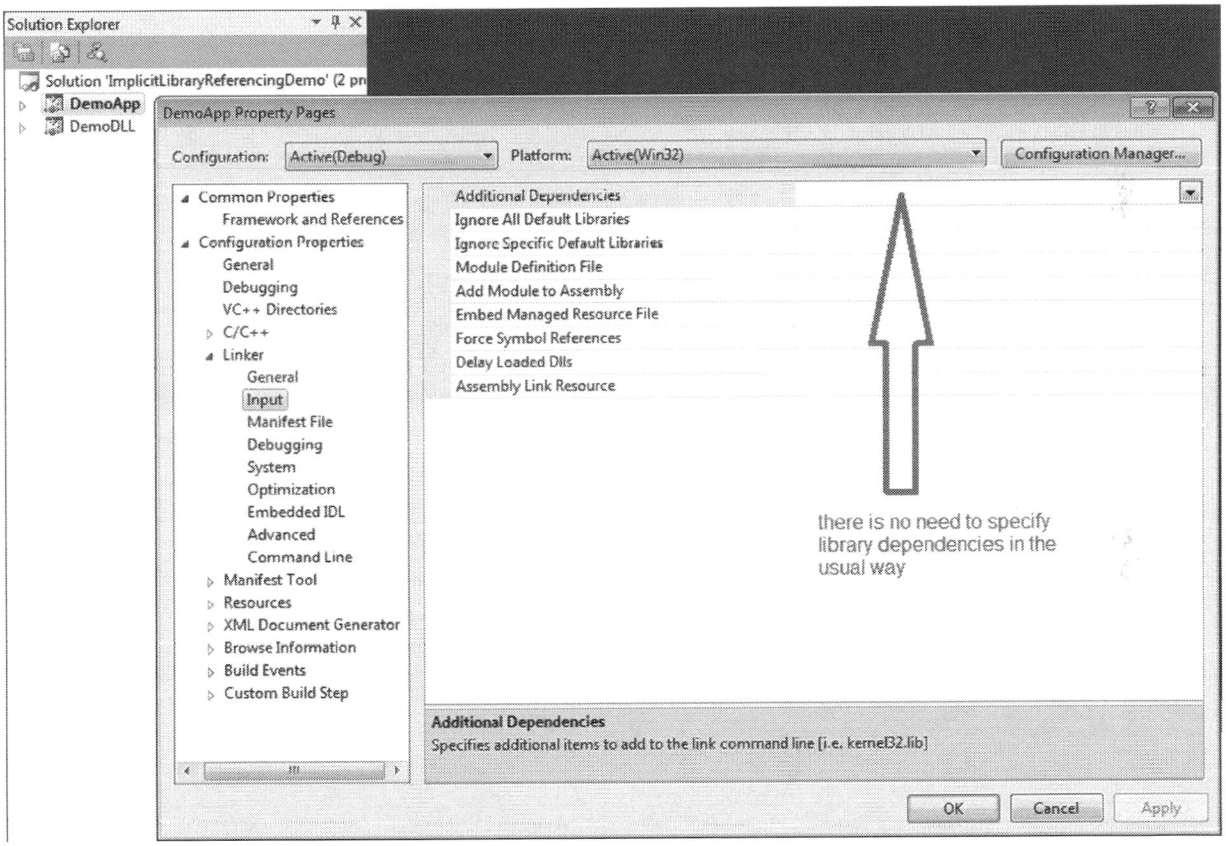

Figure 7-5. *In this method, you don't need to specify the library dependency directly*

Instead, it will be sufficient to set the client binary project to reference the dependency library project. Visit the Common Properties ➤ Frameworks and References tab (Figure 7-6).

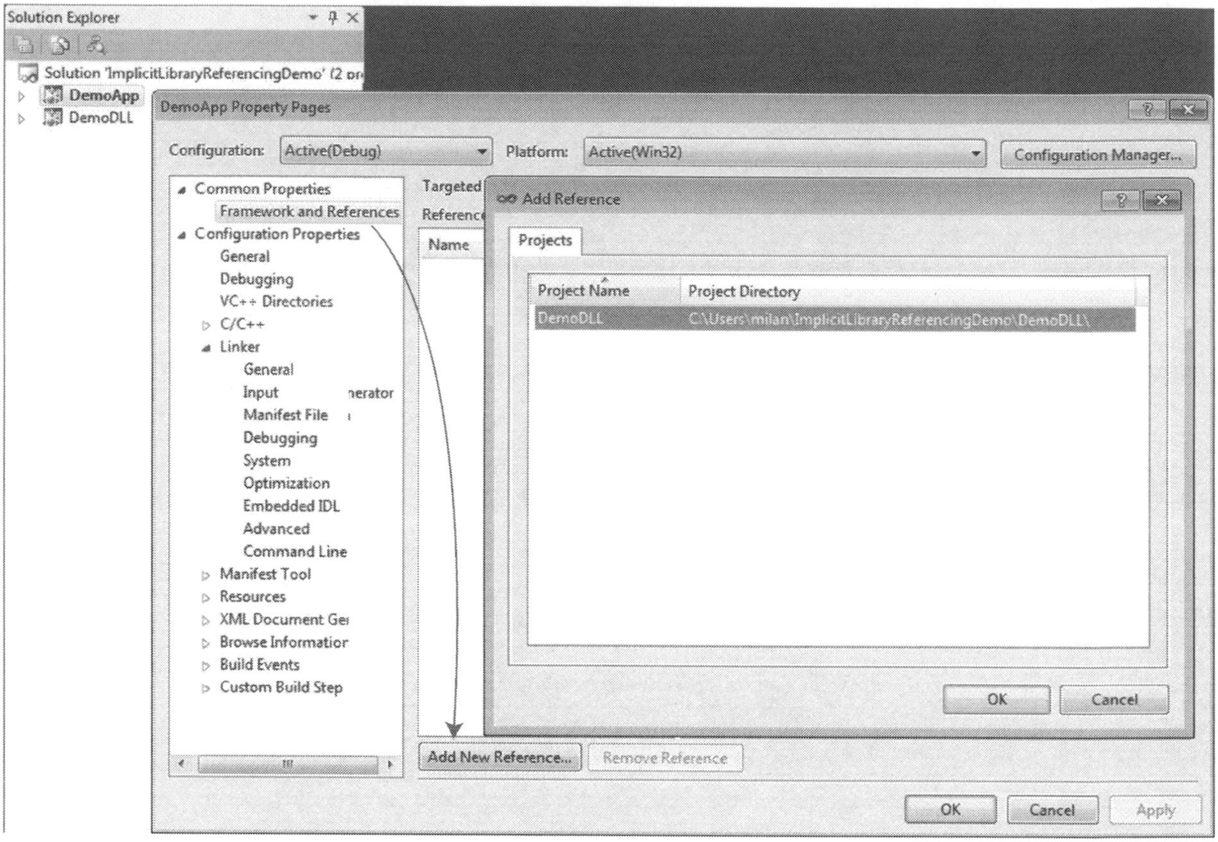

Figure 7-6. *Adding a reference to the dependency library project*

Figure 7-7 shows the referencing the dependency library project completed.

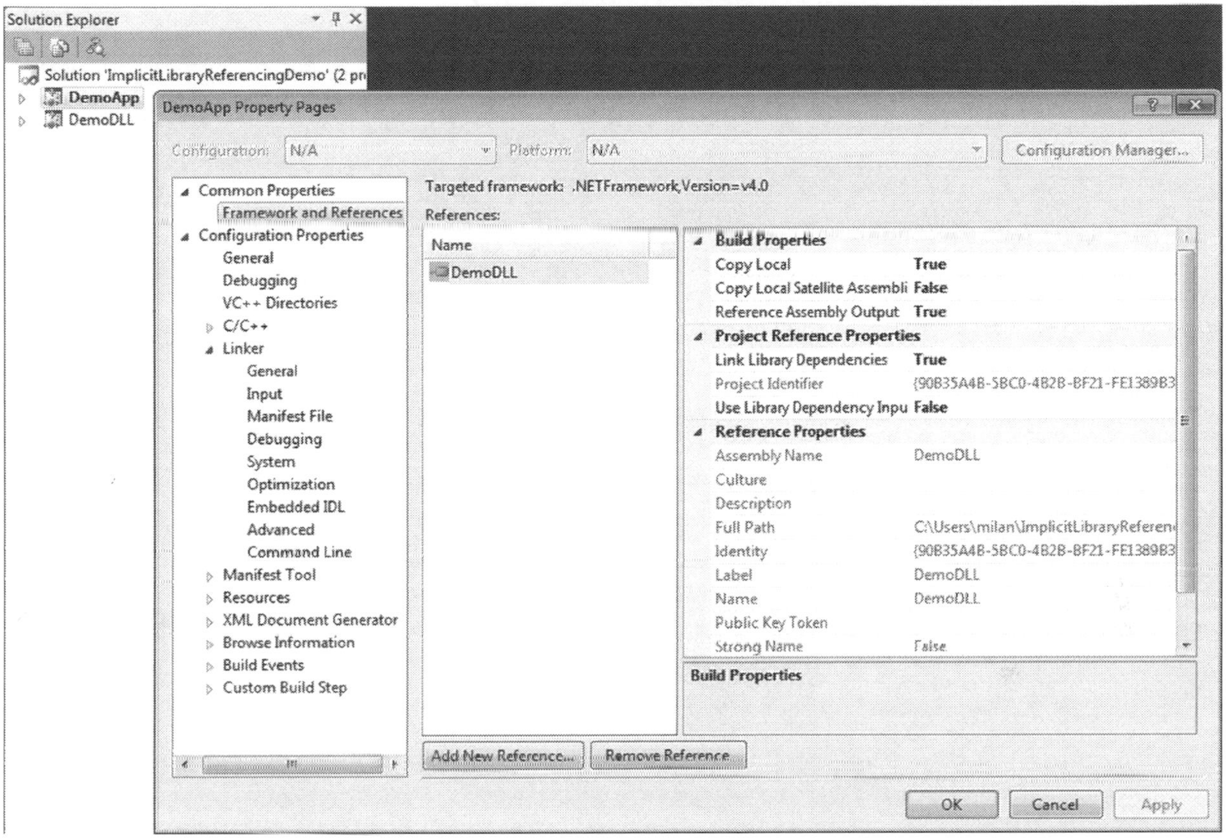

Figure 7-7. *Referencing the dependency library project is completed*

The ultimate result will be that the build time path of the required DLL is passed into the linker's command line, as shown in Figure 7-8.

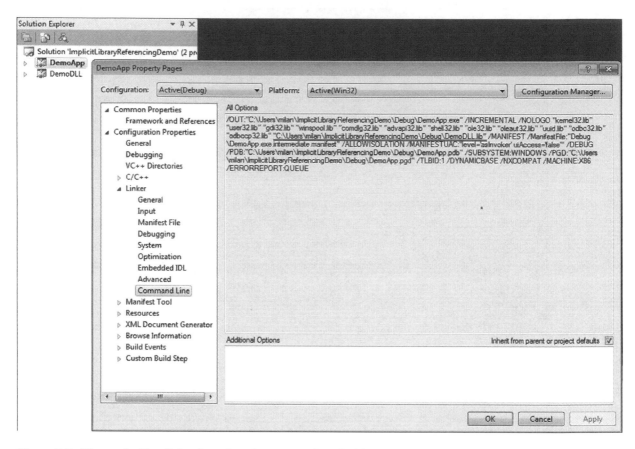

Figure 7-8. *The result of implicit referencing: the exact path to the library is passed to the linker*

Runtime Dynamic Library Location Rules

The loader needs to know the exact location of the dynamic library's binary file in order to open, read, and load it into the process. The variety of the dynamic libraries that may be needed for a program to run is vast, ranging from the always needed system libraries, all the way to the custom, proprietary, project-specific libraries.

From the programmer's perspective, the hardcoding of the paths of each and every dynamic library's path seems plain wrong. It would make much more sense if a programmer could just provide the dynamic library filename, and the operating system would somehow know where to look for the library.

All major operating systems have recognized the need to implement such a mechanism, which would be capable of searching and finding the dynamic library at runtime based upon the library filename provided by the program. Not only a set of predetermined library locations have been defined, but also the search order has been defined, specifying where the operating system will look first.

Finally, knowing the runtime location of the dynamic library is equally important regardless of whether the dynamic library is loaded statically-aware or at runtime.

Linux Runtime Dynamic Library Location Rules

The algorithm for searching for the dynamic library at runtime is governed by the following set of rules, listed in the higher order of priority.

Preloaded Libraries

The unquestionable highest priority above any library search is reserved for the libraries specified for preloading, as the loader first loads these libraries, and then starts searching for the others. There are two ways of specifying the preloaded libraries:

- By setting the **LD_PRELOAD** environment variable.

  ```
  export LD_PRELOAD=/home/milan/project/libs/libmilan.so:$LD_PRELOAD
  ```

- Through the */etc/ld.so.preload* file.

  This file contains a whitespace-separated list of ELF shared libraries to be loaded before the program.

Specifying the preloaded libraries is not the standard design norm. Instead, it is used in special scenarios, such as design stress testing, diagnostics, and emergency patching of the original code.

In the diagnostic scenarios, you can quickly create a custom version of a standard function, lace it with the debug prints, and build a shared library whose preloading will effectively replace the dynamic library that is standardly providing such function.

After the loading of the libraries indicated for preloading is finished, the search for other libraries listed as dependencies begins. It follows an elaborate set of rules, whose complete list (arranged from the topmost priority method downwards) is explained in the following sections.

rpath

From the very early days, the ELF format featured the DT_RPATH field used to store the ASCII string carrying the search path details relevant for the binary. For example, if executable XYZ depends on the runtime presence of the dynamic library ABC, than XYZ might carry in its DT_RPATH the string specifying the path at which the library ABC could be found at runtime.

This feature clearly represented a nice step ahead in allowing the programmer to establish tighter control over the deployment issues, most notably to avoid the broad scale of possible mismatches between the versions of intended vs. available libraries.

The information carried by the DT_RPATH field of the executable XYZ would be ultimately read out at runtime by the loader. An important detail to remember is that the path from which the loader is started plays a role in interpreting the DT_RPATH information. Most notably, in the cases where the DT_RPATH carries a relative path, it will be interpreted not relative to the location of library XYZ but instead relative to the path from which the loader (i.e., application) is started. As good as it is, the concept of rpath has undergone certain modifications.

According to web sources, somewhere around 1999 when version 6 of the C runtime library was in the process of superseding version 5, certain disadvantages of rpath had been noticed, and it was mostly replaced by a very similar field called runpath (DT_RUNPATH) of the ELF binary file format.

Nowadays, both rpath and runpath are available, but runpath is given higher regard in the runtime search priority list. Only in the absence of its younger sibling runpath (DT_RUNPATH field), the rpath (DT_RPATH field) remains the search path information of the highest priority for the Linux loader. If, however, the runpath (DT_RUNPATH) field of the ELF binary is non-empty, the rpath is ignored.

The rpath is typically set by passing the linker the -R or -rpath flag immediately followed by the path you want to assign as runpath. Additionally, by convention, whenever the linker is indirectly invoked (i.e., by calling gcc or g++), linker flags need to be prepended by the -Wl, prefix (i.e "minus Wl comma"):

```
$ gcc -Wl,-R/home/milan/projects/ -lmilanlibrary
       ^  ^         ^
       |  |         |
       |  |         actual rpath value
       |  |
       |  run path linker flag
       |
       -Wl, prefix required when invoking linker
       indirectly, through gcc instead of
       directly invoking ld
```

Alternatively, the rpath may be set by specifying the LD_RUN_PATH environment variable:

```
$ export LD_RUN_PATH=/home/milan/projects:$LD_RUN_PATH
```

Finally, the rpath of the binary may be modified after the fact, by running the chrpath utility program. One notable drawback of the chrpath is that it can't modify the rpath beyond its already existing string length. More precisely, chrpath can modify and delete/empty the DT_RPATH field, but cannot insert it or extend it to a longer string.

The way to examine the binary file for the value of the DT_RPATH field is to examine the binary's ELF header (such as running readelf -d or objdump -f).

LD_LIBRARY_PATH Environment Variable

It was very early on during the development of library search path concept when a need was recognized for a kind of a temporary, quick-and-dirty yet effective mechanism that could be used by developers to experiment and test their designs. The need was addressed by providing the mechanism in which a specific environment variable (LD_LIBRARY_PATH) would be used to satisfy these needs.

When the rpath (DT_RPATH) value is not set, this path supplied this way is used as the highest priority search path information.

■ **Note** In this priority scheme, there is an uneven battle between the value embedded in the binary file and the environment variable. If things stayed the same, the presence of rpath in the binary would make troubleshooting the problems with a third-party software product impossible. Fortunately, the new priority scheme addressed this problem by recognizing that the rpath is too dictatorial and by providing a way to temporarily override its settings. The rpath's younger sibling runpath is given the power to silence the rogue and authoritarian rpath, in which case LD_LIBRARY_PATH has a chance of temporarily getting the highest priority treatment.

The syntax of setting up the LD_LIBRARY_PATH is identical to the syntax of setting any kind of path variable. It can be done in the particular shell instance by typing for example something like this:

```
$ export LD_LIBRARY_PATH=/home/milan/projects:$LD_LIBRARY_PATH
```

Once again, the use of this mechanism should be reserved for the experimentation purposes. The production versions of the software products should not rely on this mechanism at all.

runpath

The runpath concept follows the same principle as the rpath. It is the field (DT_RUNPATH) of the ELF binary format, which can be set at build time to point to the path where the dynamic library should look. As opposed to the rpath, whose authority is unquestionable, the runpath is designed to be lenient to the urgent needs of LD_LIBRARY_PATH mechanism.

The runpath is set in a way very similar to how the rpath is set. In addition to passing the -R or -rpath linker flag, an additional --enable-new-dtags linker flag needs to be used. As already explained in the case of rpath, whenever the linker is called indirectly, through calling gcc (or g++) instead of invoking directly ld, by convention the linker flags need to be prepended by the -Wl, prefix:

```
$ gcc  Wl, R/home/milan/projects/ -Wl,--enable-new-dtags -lmilanlibrary
       ^   ^     ^                  ^
       |   |     |                  |
       |   |     actual rpath value both rpath and runpath set
       |   |                        to the same string value
       |   run path linker flag
       |
       -Wl, prefix required when invoking linker
       indirectly, through gcc instead of
       directly invoking ld
```

As a rule, whenever the runpath is specified, the linker sets both rpath and runpath to the same value.

The way to examine the binary file for the value of the DT_RUNPATH field is to examine the binary's ELF header (such as running readelf -h or objdump -f).

From the priority standpoint, whenever DT_RUNPATH contains a non-empty string, the DT_RPATH field is ignored by the loader. This way the dictatorial power of rpath is subdued and the will of LD_LIBRARY_PATH gets the chance of being respected when it is really needed.

The useful utility program patchelf is capable of modifying the runpath field of the binary file. At the present moment, it is not the part of the official repository, but its source code and the simple manual can be found at http://nixos.org/patchelf.html. Compiling the binary is fairly simple. The following example illustrates the patchelf use:

```
$ patchelf --set-rpath <one or more paths> <executable>
                        ^
                        |
                multiple paths can be defined,
                separated by a colon (:)
```

■ **Note** Even though the patchelf documentation mentions rpath, the patchelf in fact acts on the runpath field.

---

ldconfig Cache

One of the standard code deploy procedures is based on running the Linux ldconfig utility (http://linux.die.net/man/8/ldconfig). Running the ldconfig utility is usually one of the last steps during the standard package installation procedure, which typically requires passing the path to a folder containing libraries as input argument. The result is that ldconfig inserts the specified folder path to the list of dynamic library search folders maintained in the /etc/ld.so.conf file. At the same token, the newly added folder path is scanned for dynamic libraries, the result of which is that the filenames of found libraries get added to the list of libraries' filenames maintained in the /etc/ld.so.cache file. For example, the examination of my development Ubuntu machine reveals the contents of /etc/ld.so.conf file in Figure 7-9.

```
milan@milan$ cat /etc/ld.so.conf
include /etc/ld.so.conf.d/*.conf

milan@milan$ ls -alg /etc/ld.so.conf.d/
total 24
drwxr-xr-x   2 root  4096 Aug 17  2012 .
drwxr-xr-x 131 root 12288 Feb  5 16:09 ..
lrwxrwxrwx   1 root    40 Feb  5 15:14 i386-linux-gnu_GL.conf -> /etc/alternativ
es/i386-linux-gnu_gl_conf
-rw-r--r--   1 root   108 Apr 19  2012 i686-linux-gnu.conf
-rw-r--r--   1 root    44 Apr 19  2012 libc.conf
milan@milan-ub-1204-32-lts:~$ cat /etc/ld.so.conf.d/*
/usr/lib/i386-linux-gnu/mesa
# Multiarch support
/lib/i386-linux-gnu
/usr/lib/i386-linux-gnu
/lib/i686-linux-gnu
/usr/lib/i686-linux-gnu
# libc default configuration
/usr/local/lib
milan@milan$
```

Figure 7-9. *The contents of /etc/ld.so.conf file*

When `ldconfig` prescans all the directories listed in the /etc/ld.so.conf file, it finds gazillion dynamic libraries whose filenames it keeps in the /etc/ld.so.cache file (only a small part of which is shown in Figure 7-10).

```
milan@milan$ cat /etc/ld.so.cache
                    o
                    o
                    o
                    o
.so.0libSDL-1.2.so.0/usr/lib/i386-linux-gnu/libSDL-1.2.so.0libQtXmlPatterns.so.4
/usr/lib/i386-linux-gnu/libQtXmlPatterns.so.4libQtXml.so.4/usr/lib/i386-linux-gn
u/libQtXml.so.4libQtSvg.so.4/usr/lib/i386-linux-gnu/libQtSvg.so.4libQtSql.so.4/u
sr/lib/i386-linux-gnu/libQtSql.so.4libQtScript.so.4/usr/lib/i386-linux-gnu/libQt
Script.so.4libQtOpenGL.so.4/usr/lib/i386-linux-gnu/libQtOpenGL.so.4libQtNetwork.
so.4/usr/lib/i386-linux-gnu/libQtNetwork.so.4libQtGui.so.4/usr/lib/i386-linux-gn
u/libQtGui.so.4libQtGConf.so.1/usr/lib/libQtGConf.so.1libQtDee.so.2/usr/lib/libQ
tDee.so.2libQtDeclarative.so.4/usr/lib/i386-linux-gnu/libQtDeclarative.so.4libQt
DBus.so.4/usr/lib/i386-linux-gnu/libQtDBus.so.4libQtCore.so.4/usr/lib/i386-linux
-gnu/libQtCore.so.4libQtCLucene.so.4/usr/lib/i386-linux-gnu/libQtCLucene.so.4lib
QtBamf.so.1/usr/lib/libQtBamf.so.1libORBitCosNaming-2.so.0/usr/lib/i386-linux-gn
u/libORBitCosNaming-2.so.0libORBit-2.so.0/usr/lib/i386-linux-gnu/libORBit-2.so.0
libORBit-imodule-2.so.0/usr/lib/i386-linux-gnu/libORBit-imodule-2.so.0libLLVM-3.
0.so.1/usr/lib/i386-linux-gnu/libLLVM-3.0.so.1libI810XvMC.so.1/usr/lib/libI810Xv
MC.so.1libIntelXvMC.so.1/usr/lib/libIntelXvMC.so.1libIDL-2.so.0/usr/lib/i386-lin
ux-gnu/libIDL-2.so.0libICE.so.6/usr/lib/i386-linux-gnu/libICE.so.6libGeoIPUpdate
.so.0/usr/lib/libGeoIPUpdate.so.0libGeoIP.so.1/usr/lib/libGeoIP.so.1libGLU.so.1/
usr/lib/i386-linux-gnu/libGLU.so.1libGLEWmx.so.1.6/usr/lib/i386-linux-gnu/libGLE
Wmx.so.1.6libGLEW.so.1.6/usr/lib/i386-linux-gnu/libGLEW.so.1.6libGL.so.1/usr/lib
/i386-linux-gnu/mesa/libGL.so.1libFS.so.6/usr/lib/libFS.so.6libFLAC.so.8/usr/lib
/i386-linux-gnu/libFLAC.so.8libFLAC++.so.6/usr/lib/i386-linux-gnu/libFLAC++.so.6
libBrokenLocale.so.1/lib/i386-linux-gnu/libBrokenLocale.so.1libBrokenLocale.so/u
sr/lib/i386-linux-gnu/libBrokenLocale.sold-linux.so.2/lib/i386-linux-gnu/ld-linu
x.so.2ld-linux.so.2/lib/ld-linux.so.2milan@milan$
```

Figure 7-10. *The contents (small part) of the /etc/ld.so.cache file*

▨ **Note**   Some of the libraries referenced by the /etc/ld.so.conf file may reside in the so-called trusted library paths. If the -z nodeflib linker flag was used when building the executable, the libraries found in the OS trusted library paths will be ignored during the library search.

The Default Library Paths (/lib and /usr/lib)

The paths /lib and /usr/lib are the two default locations where Linux OS keeps its dynamic libraries. The third party programs designed to be used with superuser privileges and/or to be available to all users typically deploy their dynamic library into one of these two places.

Please notice that the path /usr/**local**/lib does *not* belong to this category. Of course, nothing prevents you from adding to the priority list by using one of the previously described mechanisms.

▨ **Note**   If the executable was linked with the -z nodeflib linker flag, all the libraries found in the OS trusted library paths will be ignored during the library search.

Priority Scheme Summary

In summary, the priority scheme has the following two operating versions.
When RUNPATH field is specified (i.e. DT_RUNPATH is non-empty)

1.   LD_LIBRARY_PATH

2.   runpath (DT_RUNPATH field)

3.   ld.so.cache

4.   default library paths (/lib and /usr/lib)

In the absence of RUNPATH (i.e. DT_RUNPATH is empty string)

1.   RPATH of the loaded binary, followed by the RPATH of the binary, which loads it all the way up to either the executable or the dynamic library which loads all of them

2.   LD_LIBRARY_PATH

3.   ld.so.cache

4.   default library paths (/lib and /usr/lib)

For more details on this particular topic, please check the Linux loader man page (http://linux.die.net/man/1/ld).

Windows Runtime Dynamic Library Location Rules

In the simplest, popular, most widespread knowledge about the topic, the following two locations are used the most as the favorite paths to deploy the DLL needed at runtime:

- The very same path in which the application binary file resides

- One of the system DLL folders (such as C:\Windows\System or C:\Windows\System32)

However, this is not where the story ends. The Windows runtime dynamic library search priority schemes are far more sophisticated as the following factors play a role in the priority scheme:

- The Windows Store applications (Windows 8) have the different set of rules than the Windows Desktop Applications.

- Whether the DLL of the same name is already loaded in the memory.

- Whether the DLL belongs to the group of known DLLs for the given version of Windows OS.

For more precise and up-to-date information, it makes the most sense to check the official Microsoft documentation about this topic, currently residing at http://msdn.microsoft.com/en-us/library/windows/desktop/ms682586(v=vs.85).aspx.

Linux Demo of Build Time and Runtime Conventions

The following example illustrates the positive effects of tightly following the -L and -R conventions. The project used in this example is comprised of the dynamic library project and its test application project. In order to demonstrate the importance of applying the -L convention, the two demo applications are created. The first one, aptly named testApp_withMinusL demonstrates the positive effects of using the -L linker flag. The other one (testApp_withoutMinusL) demonstrates what kind of troubles may happen if the -L convention is not followed.

Both applications also rely on rpath option to specify the runtime location of the required dynamic library. The dynamic library's project folder and the apps' project folder are structured like Figure 7-11.

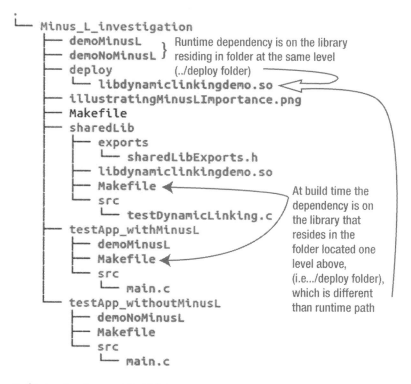

```
.
└── Minus_L_investigation
    ├── demoMinusL          ⎫ Runtime dependency is on the library
    ├── demoNoMinusL        ⎬ residing in folder at the same level
    ├── deploy                (../deploy folder)
    │   └── libdynamiclinkingdemo.so ⇐
    ├── illustratingMinusLImportance.png
    ├── Makefile
    ├── sharedLib
    │   ├── exports
    │   │   └── sharedLibExports.h
    │   ├── libdynamiclinkingdemo.so
    │   ├── Makefile ◄
    │   └── src
    │       └── testDynamicLinking.c
    ├── testApp_withMinusL
    │   ├── demoMinusL
    │   ├── Makefile ◄
    │   └── src
    │       └── main.c
    └── testApp_withoutMinusL
        ├── demoNoMinusL
        ├── Makefile
        └── src
            └── main.c
```

At build time the dependency is on the library that resides in the folder located one level above, (i.e.../deploy folder), which is different than runtime path

```
9 directories, 15 files
```

Figure 7-11. *The folder structure of project designed to illustrate the benefits of strictly following the –L –l conventions*

The Makefile of the application not relying on the -L convention looks like Listing 7-1.

Listing 7-1. Makefile Not Relying on the –L Convention

```
# Import includes
COMMON_INCLUDES  = -I../sharedLib/exports/

# Sources/objects
SRC_PATH         = ./src
OBJECTS          = $(SRC_PATH)/main.o

# Libraries
SYSLIBRARIES     =                 \
                   -lpthread \
                   -lm           \
                   -ldl

DEMOLIB_PATH     = ../deploy
# specifying full or partial path may backfire at runtime !!!
DEMO_LIBRARY     = ../deploy/libdynamiclinkingdemo.so
LIBS             = $(SYSLIBRARIES) $(DEMO_LIBRARY) -Wl,-Bdynamic

# Outputs
FXECUTABLE       = demoNoMinusL

# Compiler
INCLUDES         = $(COMMON_INCLUDES)
DEBUG_CFLAGS     = -Wall -g -O0
RELEASE_CFLAGS   = -Wall -O2

ifeq ($(DEBUG), 1)
CFLAGS           = $(DEBUG_CFLAGS) $(INCLUDES)
else
CFLAGS           = $(RELEASE_CFLAGS) $(INCLUDES)
Endif

COMPILE          = g++ $(CFLAGS)

# Linker
RUNTIME_LIB_PATH = -Wl,-R$(DEMOLIB_PATH)
LINK             = g++

# Build procedures/target descriptions
default: $(EXECUTABLE)
%.o: %.c
        $(COMPILE) -c $< -o $@
$(EXECUTABLE): $(OBJECTS)
        $(LINK) $(OBJECTS) $(LIBS) $(RUNTIME_LIB_PATH) -o $(EXECUTABLE)
clean:
        rm $(OBJECTS) $(EXECUTABLE)
deploy:
        make clean; make; patchelf --set-rpath ../deploy:./deploy $(EXECUTABLE);\
        cp $(EXECUTABLE) ../;
```

The Makefile of the application following the -L convention looks like Listing 7-2.

Listing 7-2. Makefile Following the –L Convention

```
# Import includes
COMMON_INCLUDES  = -I../sharedLib/exports/

# Sources/objects
SRC_PATH         = ./src
OBJECTS          = $(SRC_PATH)/main.o

# Libraries
SYSLIBRARIES     =                \
                    -lpthread \
                    -lm         \
                    -ldl

SHLIB_BUILD_PATH = ../sharedLib
DEMO_LIBRARY     = -L$(SHLIB_BUILD_PATH) -ldynamiclinkingdemo
LIBS             = $(SYSLIBRARIES) $(DEMO_LIBRARY) -Wl,-Bdynamic

# Outputs
EXECUTABLE       = demoMinusL

# Compiler
INCLUDES         = $(COMMON_INCLUDES)
DEBUG_CFLAGS     = -Wall -g -O0
RELEASE_CFLAGS   = -Wall -O2

ifeq ($(DEBUG), 1)
CFLAGS           = $(DEBUG_CFLAGS) $(INCLUDES)
else
CFLAGS           = $(RELEASE_CFLAGS) $(INCLUDES)
endif

COMPILE          = g++ $(CFLAGS)

# Linker
DEMOLIB_PATH     = ../deploy
RUNTIME_LIB_PATH = -Wl,-R$(DEMOLIB_PATH)

LINK             = g++

# Build procedures/target descriptions
default: $(EXECUTABLE)
%.o: %.c
        $(COMPILE) -c $< -o $@
$(EXECUTABLE): $(OBJECTS)
        $(LINK) $(OBJECTS) $(LIBS) $(RUNTIME_LIB_PATH) -o $(EXECUTABLE)
clean:
        rm $(OBJECTS) $(EXECUTABLE)
deploy:
        make clean; make; patchelf --set-rpath ../deploy:./deploy $(EXECUTABLE);\
        cp $(EXECUTABLE) ../;
```

When the process of building the dynamic library is completed, its binary is deployed to the deploy folder, which resides the two levels of depth above the folder in which the application Makefile resides. Hence, the build-time path needs to be specified as ../deploy/libdynamiclinkingdemo.so.

Figure 7-12 illustrates the advantage of adhering to the -L convention: the immunity of the program to the change of the runtime library paths.

```
/Minus_L_investigation$ ls -alg

2 21:30 .
2 21:34 
2 21:33 demoMinusL
2 21:33 demoNoMinusL
2 21:30 deploy
2 21:15 Makefile
2 21:33 sharedLib
2 21:33 testApp_withMinusL
2 21:33 testApp_withoutMinusL
/Minus_L_investigation$
/Minus_L_investigation$
/Minus_L_investigation$
/Minus_L_investigation$
/Minus_L_investigation$ ldd demoMinusL
      linux-gate.so.1 =>  (0xb77d9000)
      libdynamiclinkingdemo.so => ./deploy/libdynamiclinkingdemo.so (0xb77d4000)
      libc.so.6 => /lib/i386-linux-gnu/libc.so.6 (0xb7612000)
      /lib/ld-linux.so.2 (0xb77da000)
/Minus_L_investigation$
/Minus_L_investigation$
/Minus_L_investigation$
/Minus_L_investigation$
/Minus_L_investigation$ ldd demoNoMinusL
      linux-gate.so.1 =>  (0xb7700000)
      ../deploy/libdynamiclinkingdemo.so => not found
      libc.so.6 => /lib/i386-linux-gnu/libc.so.6 (0xb753c000)
      /lib/ld-linux.so.2 (0xb7701000)
/Minus_L_investigation$
/Minus_L_investigation$
/Minus_L_investigation$
/Minus_L_investigation$ mkdir ../deploy
/Minus_L_investigation$ cp ./deploy/libdynamiclinkingdemo.so ../deploy/
/Minus_L_investigation$
/Minus_L_investigation$
/Minus_L_investigation$
/Minus_L_investigation$ ldd demoNoMinusL
      linux-gate.so.1 =>  (0xb77d1000)
      ../deploy/libdynamiclinkingdemo.so (0xb77cc000)
      libc.so.6 => /lib/i386-linux-gnu/libc.so.6 (0xb760a000)
      /lib/ld-linux.so.2 (0xb77d2000)
/Minus_L_investigation$
```

Library specified as -L<path> -l <name> may be handled neatly in both linking as well as at runtime (where its name may be cleanly combined with rpath.)

Library specified without requires worrying about maintaining the relative paths.

Figure 7-12. *The benefit of carefully following the –L –l conventions. Following the convention typically means being worry free at runtime*

When the build time library path was specified with the -L option, the library name is effectively separated from the path and as such imprinted into the client binary file. When the time comes to conduct the runtime search, the imprinted name (i.e., not the path plus name, but solely the library name!) fits really well with the implementation of the runtime search algorithm.

■ ■ ■

## CHAPTER 8

Designing Dynamic Libraries: Advanced Topics

The purpose of this chapter is to discuss the details of the dynamic linking process. A crucial factor in this process is the memory mapping concept. Basically, it allows a dynamic library that is already loaded in the memory map of a running process to be mapped into the memory map of another process running concurrently.

The important rule of dynamic linking is that different processes do share the code segment of the dynamic library, but do not share the data segments. Each of the processes loading the dynamic library is expected to provide its own copy of the data on which the dynamic library code operates (i.e., the library's data segment). Following the culinary analogy, several chefs in several restaurants can concurrently use the same book of recipes (instructions). It is very likely, however, that different chefs will use different recipes from the same book. Also, it is given that the dishes prepared by the recipes from the same cook book will be served to different customers. Obviously, despite the fact that the chefs read from the same recipe book, they each should use their own set of dishes and kitchen utensils. Otherwise, it would be a huge mess.

As great and simple as the whole story looks now, several technical problems needed to be solved along the way. Let's take a closer look.

Why Resolved Memory Addresses Are a Must

Before going into the details of the technical problems encountered during the design of dynamic linking implementations, it is worth reiterating a few simple facts rooted in the domain of assembly language and machine instructions that ultimately determine so many other details.

Namely, certain groups of instructions expect that the address of the operand in memory be known at runtime. In general, the following two groups of instructions strictly require the precisely calculated addresses:

- Data access instructions (mov, etc.) require the address of the operand in memory. For example, in order to access a data variable, the mov assembler instruction of X86 architecture requires the absolute memory address of the variable in order to transfer the data between the memory and a CPU register.

    The following sequence of assembly instructions is used to increment a variable stored in memory:

```
mov eax, ds:0xBFD10000 ; load the variable from address 0xBFD10000 to register eax
add eax, 0x1           ; increment the loaded value
mov ds:0xBFD10000, eax ; store the result back to the address 0xBFD10000
```

- Subroutine calls (call, jmp, etc.) require the address of function in code segment. For example, in order to call a function, the call instruction must be supplied with the code segment memory address of the function entry point.

  The following sequence of assembly instructions performs the actual function call:

  ```
  call 0x0A120034 ; calling function whose entry point resides at address 0x0A120034
  ```

  which is equivalent to

  ```
  push eip + 2    ; return address is current address + size of two instructions
  jmp 0x0A120034  ; jumping to the address of my_function
  ```

For things to be somewhat easier, there are scenarios in which merely a relative offset plays a role. The addresses of static variables as well as the entry points of functions of local scope (both declared by using the static keyword in the sense of C programming language) may be resolved by knowing only the relative offset from the instructions that reference them. Both data access and/or subroutine call assembler instructions do have the flavors that require the relative offset instead of the absolute address. This, however, does not remove the overall problem; it only diminishes it to an extent.

General Problem of Resolving References

Let's consider the simplest possible case in which the executable (application) is the client binary that loads a single dynamic library. The following set of known facts describes the working scenario:

- A fixed, predetermined part of the process memory map blueprint is provided by the executable binary.

- Once the dynamic loading is completed, the dynamic library becomes a legitimate part of the process.

- The connection between the executable and the dynamic library naturally happens by the executable calling one or more functions implemented and properly exported by a dynamic library.

Here comes the interesting part.

The process of loading the library into the process memory map begins by translating the library segment's address range to a new location. In general, the address range where the dynamic library will be loaded is not known in advance. Instead, it is determined at load time by the internal algorithm of the loader module.

The level of indeterminism in this scenario is only slightly diminished by the fact that the executable format stipulates the address range where the dynamic library may be loaded. However, the stipulated range of allowed addresses is fairly broad, as it was designed to accommodate many dynamic libraries loaded at the same time. This clearly does not help much in guessing where exactly the dynamic library will eventually be loaded.

The process of address translation (illustrated in Figure 8-1) that happens during the dynamic library loading is the crucial problem of dynamic linking, which makes the whole concept rather complex.

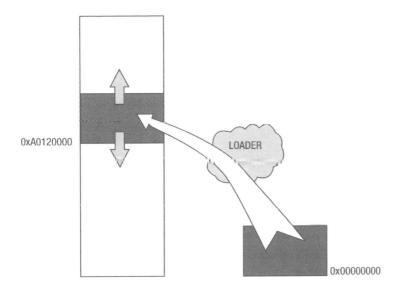

Figure 8-1. *Address translation inevitably happens as the loader tries to find a place for the dynamic library in the process memory map*

What exactly is the problem with the address translation?

The address translation is not the problem per se. You've seen in previous chapters that the linker routinely performs this simple operation when trying to tile the object files together into the process memory map. However, it is very important which module performs the address translation.

More specifically, there are substantial differences in the scenario in which *the linker* performs the address translation from the scenario in which *the loader* does the same thing.

- When performing the address translation, the linker in general has a "clean slate/virgin snow" situation. None of the object files taken in by the linker during the tiling process have any of the references resolved. This gives the linker a huge degree of freedom to juggle the object files when trying to find the right place for them. Upon completing the initial placement of object files, the linker scans the list of unresolved references, resolves them, and stamps the correct addresses into the assembly instructions.

- The loader, on the other hand, operates within significantly different circumstances. It takes as input the dynamic library binary, which already passed the complete building process, and has resolved all the references. In other words, all of the assembler instructions are stamped with the correct addresses.

  In the particular cases in which the linker imprinted the absolute addresses into the assembler instructions, *the address translation performed by the loader makes imprinted addresses completely meaningless.* Executing such fundamentally disrupted instructions gives bogus results at best, and has a potential of being very dangerous. Obviously, the address translation performed during the dynamic loading kind of falls into the broad category of the "elephant in the china store" paradigm.

In summary, the address translation of the loader cannot be avoided, as it is inherent to the idea of dynamic loading. However, it immediately imposes a very serious kind of problem. Fortunately, even though it cannot be avoided, a few ways of dancing around it have been successfully implemented.

Which Symbols Are Likely to Suffer from Address Translation?

It is almost a no-brainer that the functions and variables declared *static* (in the C language sense, as in relevant only for the file in which they reside) are out of danger. Indeed, since only the nearby instructions need to access these symbols, all the accesses can be implemented by supplying the relative address offsets.

What is the situation with the functions and variables that are not declared static?

As it turns out, not being declared static still does not mean that such function or variable will inevitably be doomed to suffer from the address translation.

In fact, *only the functions and variables whose symbols are **exported*** by the dynamic library are guaranteed to suffer from the negative effects of the address translation. In fact, when the linker knows that a certain symbol is exported, it implements all the accesses via the absolute addresses. The address translation then renders such instructions unusable.

The code example analyzed in Appendix A illustrates this point, in which two non-static variables are featured in the code, of which only one is exported by the dynamic library. As the analysis shows, the exported variable is the one that gets affected by the dynamic loading address translation.

Problems Caused by Address Translation

There are times when address translation during dynamic loading can cause problems. Fortunately, these can be systematized into two general scenarios.

Scenario 1: Client Binary Needs to Know the Address of Dynamic Library Symbols

This is the most basic scenario, which happens when the client binary (an executable or a dynamic library) counts on symbols of a loaded dynamic library being available at runtime, but does not know what the ultimate address will be, as illustrated in Figure 8-2.

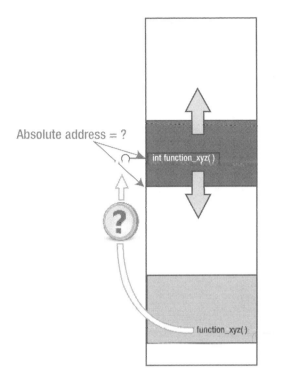

Absolute address = ?

int function_xyz( )

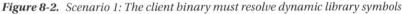

function_xyz( )

Figure 8-2. *Scenario 1: The client binary must resolve dynamic library symbols*

If you assume the usual approach where the task of resolving the symbol addresses traditionally belongs to the linker (and linker only), you are in a troubled situation. Namely, the linker already completed its task of building both the client binary as well as the library, which is being dynamically loaded.

It quickly becomes very obvious that certain "out of the box" thinking needs to be applied in order to resolve this kind of situation. The solution leads into the direction of granting a part of the linker's responsibilities of resolving the symbols to the loader.

In the new scheme of things, the new ability of the loader to take some of the linker's abilities is typically implemented as a module commonly referred to as a ***dynamic linker***.

Scenario 2: Loaded Library No Longer Knows the Addresses of Its Own Symbols

Typically, the ABI functions exported by the dynamic libraries are the well encapsulated entry points into the library's inner functionality. The typical sequence that happens at runtime is that the client binary typically calls one of the ABI methods, which in turn calls the library's internal functions, which are of no particular interest to the client binary and hence not exported.

A possible different scenario (albeit somewhat less frequently encountered) is when a dynamic library ABI function internally calls the other ABI function.

Let's assume, for example, that a dynamic library hosts a module that exports the two interface functions:

- `Initialize()`
- `Uninitialize()`

The internal execution flow of each of the two functions will very likely assume the sequence of calls of library internal functions, declared with a static scope. Calling the internal methods is typically performed by the assembler call family of instructions featuring relative addresses. The address translation does not negatively affect the implementation of call functions, as illustrated in Figure 8-3.

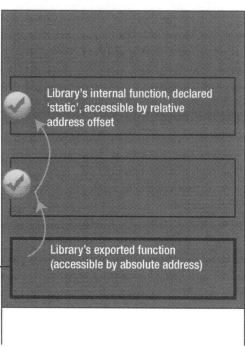

Regardless of the fact that the absolute address of ABI function is not known, the relative offsets to the library's internal static functions are sufficient to implement the call instructions.

The address translation did not cause trouble at least in this part of the overall picture.

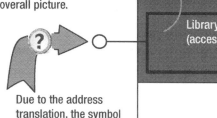

Due to the address translation, the symbol address is unknown.

Figure 8-3. *Regardless of address translation, the calls to local functions (which may may be implemented as relative jumps) can be easily resolved*

It is perfectly possible, however, that the library designers decide to provide the Reinitialize() interface function. It would neither be surprising nor wrong that this function internally first calls the Uninitialize() interface function, immediately followed by the call to the Initialize() interface function.

By being the ABI interface function, the Reinitialize() function's entry point must belong to the set of dynamic library's exported symbols. The jump instructions that refer to this function may *not* be implemented as relative jumps. Instead, the linker must implement the jump/call instructions as jumps to the absolute addresses.

Obviously, now you have an interesting kind of situation. The damaged party in this scenario is no longer only the client binary, but the loaded library as well. After the memory translation is performed by the loader, the function addresses are no longer applicable. The assembler call instructions that the linker nicely imprinted with the absolute addresses are not only meaningless, but potentially dangerous, as their jump target is no longer where it was originally planned to be, as shown in Figure 8-4.

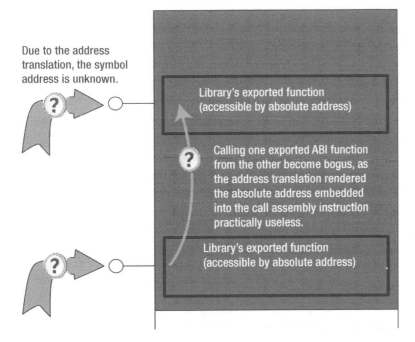

Due to the address
translation, the symbol
address is unknown.

Figure 8-4. *Scenario 2: One ABI function internally calling another suffers from unresolved references problems. Both function entry points are designated for export, which urges the compiler to implement calls as absolute jumps. Resolving the absolute addresses is not possible until the loader completes the address translation*

Again, the identical problem that you face with the ABI functions exists with the dynamic library's global scope variables.

Linker-Loader Coordination

It was recognized early on that in the dynamic linking scenario the linker can not completely solve everything that it normally solves during the building of monolithic executable.

During the initial stage of dynamic linking, the loader loads the code segment of the dynamic library into the new address range. Even though the linker legitimately finished the task of resolving the references when building the dynamic library, it is simply not enough; the address translation process rendered the absolute addresses imprinted into the assembler call instructions invalid.

The fact that the "earthquake" happens after the linker did all it could do implies that there must be "someone smart" to fix the troubles after the fact. That "someone smart" was chosen to be the loader.

Overall Strategy

Knowing all the previously described constraints, the cooperation between the linker and loader has been established according to the following set of broad guidelines:

- The linker recognizes its own limitations.

- The linker precisely estimates the damage, prepares the directives for fixing it, and embeds the directives into the binary file.

- The loader precisely follows the linker directives and applies the corrections after the address translation is completed.

Linker Recognizes Its Own Limitations

When creating a dynamic library, in addition to being witty in figuring out the relationships between the various parts of the puzzle, the linker also must be smart enough to recognize what will be disrupted as result of the code segment loading into the different range of addresses.

First, the code address range of the dynamic library memory map is zero-based, unlike with executables, in which case the linker deals with the more specific non-zero based address ranges.

Second, when recognizing that the addresses of certain symbols cannot be resolved before the load time, the linker stops trying; instead, it fills the unresolved symbols with temporary values (typically being some obviously wrong values, such as all zeros or so).

This does not mean, however, that linker has given up the quest of completing the task.

Linker Precisely Estimates the Damage, and Prepares Directives for Fixing It

It is completely possible to classify all the scenarios in which the loader address translation will render ineffective the previously resolved references. Such cases happen whenever the assembler instructions require absolute addresses. When completing the linking stage of building the dynamic library, the linker can identify such occurrences and somehow let the loader know of them.

In order to provide support for linker-loader coordination, the binary format specifications support brand new sections whose purpose is solely to provide the place for the linker to leave the directives for the loader of how to fix the damage caused by address translation happening during the dynamic loading. Furthermore, a specific simple syntax has been devised so that the linker can precisely specify to the loader the course of action to take. Such sections are called relocation sections in the binary, of which the .rel.dyn section is the oldest.

In general, the relocation directives are written into the binary by the linker, to be read later by the loader. They specify

- At which addresses the loader needs to apply some patching after laying out the final memory map of the whole process.

- What exactly the loader needs to do in order to correctly patch the unresolved addresses.

The Loader Precisely Follows the Linker Directives

The last phase belongs to the loader. It reads in the dynamic library created by the linker, reads in the loader segments (each carrying a variety of linker sections), and lays them all out into the process memory map, alongside the code belonging to the original executable.

Finally, it locates the .rel.dyn section, reads in the directives that the linker left, and according to these directives performs patching of the original dynamic library. When patching is completed, the memory map is ready for launching the process.

Obviously, the task of handling the dynamic library loading requires that some more intelligence be granted to the loader than what it needs for its basic tasks.

Tactics

In general, the exchange of information between the linker and the loader happens through the specific .rel.dyn section inserted by the linker into the body of the binary. The only question is *in which of the binary files will the linker insert the* .rel.dyn *section?*

The answer is simple: it is the squeaky wheel that gets the oil. The binary whose code section needs repair will generally carry the .rel.dyn section.

Concretely, in Scenario 1, the linker embeds the .rel.dyn section into the client binary (the executable or dynamic library whose instructions were "damaged" by the loading of a new dynamic library), as this is where the address translation of loaded libraries caused troubles. Figure 8-5 illustrates the idea.

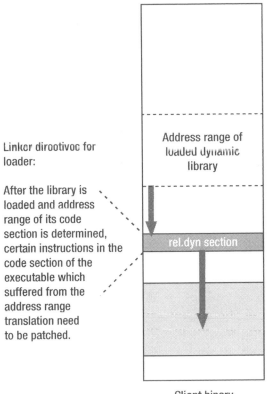

Linker directives for loader:

After the library is loaded and address range of its code section is determined, certain instructions in the code section of the executable which suffered from the address range translation need to be patched.

Figure 8-5. *In Scenario 1, the linker directives are embedded into the client binary file*

In Scenario 2, however, the linker embeds the .rel.dyn section into the binary of the loaded library, as it needs help reconstructing the coherence between the addresses and instructions that point to them (Figure 8-6).

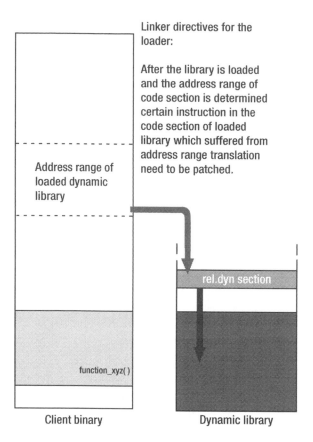

Linker directives for the loader:

After the library is loaded and the address range of code section is determined certain instruction in the code section of loaded library which suffered from address range translation need to be patched.

Address range of loaded dynamic library

rel.dyn section

function_xyz( )

Client binary

Dynamic library

Figure 8-6. *In Scenario 2, the linker directives are embedded into the dynamic library*

In this particular example, you have the simplest possible scenario in which an executable loads a dynamic library. A far more realistic case is that a dynamic library itself may load another dynamic library, which in turn may load yet another dynamic library, etc. Any of the dynamic libraries in the middle of the chain of dynamic loading may be in dual role. Scenario 1 as well as Scenario 2 may happen to be applicable to the same binary.

Linker Directives Overview

The binary format specifications typically specify in detail the syntax rules of communication between the linker and the loader. The linker directives for the loader generally tend to be very simple, yet very precise and to the point (Figure 8-7). Hence the structuring the information carried by the linker directives does not take tremendous amount of effort to implement and understand.

```
 Offset      Info     Type              Sym.Value   Sym. Name
0804a000   00000107 R_386_JUMP_SLOT    00000000    printf
0804a004   00000207 R_386_JUMP_SLOT    00000000    shlib_abi_function
0804a008   00000307 R_386_JUMP_SLOT    00000000    __gmon_start__
0804a00c   00000407 R_386_JUMP_SLOT    00000000    dl_iterate_phdr
0804a010   00000507 R_386_JUMP_SLOT    00000000    __libc_start_main
0804a014   00000607 R_386_JUMP_SLOT    00000000    putchar


 Offset      Info     Type              Sym.Value   Sym. Name
000004b8   00000008 R_386_RELATIVE
00002008   00000008 R_386_RELATIVE
000004c8   00000801 R_386_32           0000201c    shlibNonStaticAccessed
000004d0   00000801 R_386_32           0000201c    shlibNonStaticAccessed
000004ea   00000b01 R_386_32           0000200c    shlibNonStaticVariable
00001fe8   00000106 R_386_GLOB_DAT     00000000    __cxa_finalize
00001fec   00000206 R_386_GLOB_DAT     00000000    __gmon_start__
00001ff0   00000306 R_386_GLOB_DAT     00000000    _Jv_RegisterClasses
```

Figure 8-7. *Examples of linker directives*

In particular, the ELF file format carries the detailed definitions of how the linker specifies the directives for the loader. The directives are stored primarily in the .rel.dyn section as well as in a few other specialized sections (rel.plt, got, got.plt). The tools such as readelf or objdump can be used to display the contents of the directives. Figure 8-7 shows some of the examples.

The interpretation of the fields of the directive syntax is the following:

- **Offset** specifies the code section byte offset to the assembler instruction operand, which is rendered meaningless by the address translation and needs repair.

- **Info** is described by the ELF format specification as

```
#define ELF32_R_SYM(i)   ((i)>>8)
#define ELF32_R_TYPE(i)  ((unsigned char)(i))
#define ELF32_R_INFO(s,t) (((s)<<8)+(unsigned char)(t))

#define ELF64_R_SYM(i)   ((i)>>32)
#define ELF64_R_TYPE(i)  ((i)&0xffffffffL)
#define ELF64_R_INFO(s,t) (((s)<<32)+((t)&0xffffffffL))
```

where

- ELFxx_R_SYM denotes the symbol table index with respect to which the relocation must be made:

  One of the sections of the binary file carries the list of symbols. This value simply represents the index of the symbol table item that represents this particular symbol. The readelf and objdump can provide the complete list of symbols contained in the binary's symbol table.

- ELFxx_R_TYPE denotes the type of relocation to apply. A detailed description of available relocation types is shown below.

- **Type** specifies the type of action that the loader needs to perform on the assembler instruction operand in order to repair the problems caused by the address translation. The ELF binary format shown in Figure 8-8 (Figure 1-22 of the ELF specification) specifies the following relocation types.

Figure 1-22: Relocation Types

Name	Value	Field	Calculation
R_386_NONE	0	none	none
R_386_32	1	*word32*	S + A
R_386_PC32	2	*word32*	S + A - P
R_386_GOT32	3	*word32*	G + A - P
R_386_PLT32	4	*word32*	L + A - P
R_386_COPY	5	none	none
R_386_GLOB_DAT	6	*word32*	S
R_386_JMP_SLOT	7	*word32*	S
R_386_RELATIVE	8	*word32*	B + A
R_386_GOTOFF	9	*word32*	S + A - GOT
R_386_GOTPC	10	*word32*	GOT + A - P

Some relocation types have semantics beyond simple calculation.

R_386_GOT32   This relocation type computes the distance from the base of the global offset table to the symbol's global offset table entry. It additionally instructs the link editor to build a global offset table.

R_386_PLT32   This relocation type computes the address of the symbol's procedure linkage table entry and additionally instructs the link editor to build a procedure linkage table.

R_386_COPY   The link editor creates this relocation type for dynamic linking. Its offset member refers to a location in a writable segment. The symbol table index specifies a symbol that should exist both in the current object file and in a shared object. During execution, the dynamic linker copies data associated with the shared object's symbol to the location specified by the offset.

R_386_GLOB_DAT   This relocation type is used to set a global offset table entry to the address of the specified symbol. The special relocation type allows one to determine the correspondence between symbols and global offset table entries.

R_3862_JMP_SLOT   The link editor creates this relocation type for dynamic linking. Its offset member gives the location of a procedure linkage table entry. The dynamic linker modifies the procedure linkage table entry to transfer control to the designated symbol's address [see "Procedure Linkage Table" in Part 2].

R_386_RELATIVE   The link editor creates this relocation type for dynamic linking. Its offset member gives a location within a shared object that contains a value representing a relative address. The dynamic linker computes the corresponding virtual address by adding the virtual address at which the shared object was loaded to the relative address. Relocation entries for this type must specify 0 for the symbol table index.

R_386_GOTOFF   This relocation type computes the difference between a symbol's value and the address of the global offset table. It additionally instructs the link editor to build the global offset table.

R_386_GOTPC   This relocation type resembles R_386_PC32, except it uses the address of the global offset table in its calculation. The symbol referenced in this relocation normally is _GLOBAL_OFFSET_TABLE_, which additionally instructs the link editor to build the global offset table.

Figure 8-8. *Overview of linker directive types (from ELF format specification)*

- Sym.Value specifies the tentative, temporary offset within the code section (in case of functions) or within the data segment (in case of variables) where the symbol currently resides in the original binary file. It is assumed that the address translation will affect these values.

- Sym.Name specifies the human readable symbol name (function name, variable name)

Linker-Loader Coordination Implementation Techniques

Throughout the evolution of the dynamic linking concept, the two implementation techniques have been used: Load Time Relocation (LTR) and Position Independent Code (PIC).

Load Time Relocation (LTR)

Chronologically, the first implementation of the dynamic linking concept came in the form of the so-called Load Time Relocation. Broadly speaking, this technique was the first dynamic loading technique that really worked. Its immediate benefit was the ability to relieve the application binaries from the need to carry along unnecessary "luggage" (the code that handles the usual chores specific to the operating system).

The immediate benefits that the LTR concept brought was that not only the byte size of applications' binaries became substantially smaller, but also the way certain OS-specific tasks were executed became unified across the wide variety of applications.

Despite the obvious benefits that this concept brought, it had several major drawbacks. First, this technique modifies (patches) the dynamic library code with the literal values of addresses of variables and functions, meaningful only in the context of the application that loaded it first. In the context of any other application (which would very possibly feature the different process' memory map layout), the code modifications would very likely be useless, meaningless, and simply not applicable.

As a consequence, if several applications needed the services of a dynamic library at the same time, it would mean that you would have exactly that many copies of the same dynamic library in memory.

The second drawback was that a proportionally large amount of code modifications would be required. With this technique in place, the loader needs to modify/patch exactly as many places in the code that reference a certain variable or call a certain function. In cases where the application loads plenty of dynamic libraries, the load time grows to significant and noticeable initial latency during the application start.

The third drawback was that the writable code (.text) segment poses a potential security threat.

With this technique in place, the dream of loading the dynamic library into the physical memory only once, and mapping it into the plethora of different applications' memory maps different address, was not achievable.

Figure 8-9 illustrates the idea behind the Load Time Relocation concept.

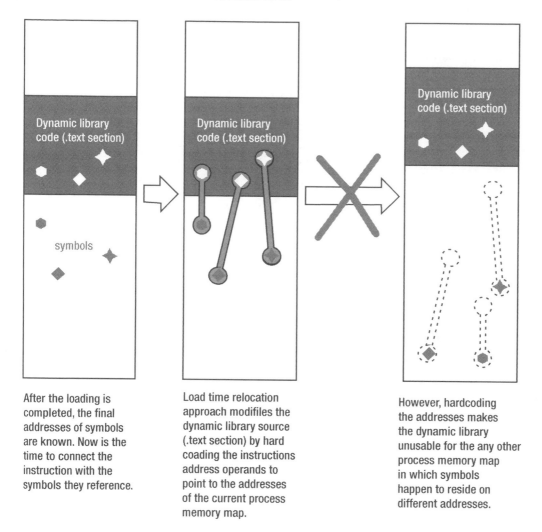

After the loading is completed, the final addresses of symbols are known. Now is the time to connect the instruction with the symbols they reference.

Load time relocation approach modifiles the dynamic library source (.text section) by hard coading the instructions address operands to point to the addresses of the current process memory map.

However, hardcoding the addresses makes the dynamic library unusable for the any other process memory map in which symbols happen to reside on different addresses.

Figure 8-9. *LTR concept and its limitations*

All the drawbacks were addressed with the design of the newer and in many aspects superior Position Independent Code (PIC) approach, which quickly became the prevalent choice of linking techniques.

Position Independent Code (PIC)

The limitations of the Load Time Relocation scheme have been addressed in the next implementation of the dynamic linking, the technique known as Position Independent Code (Figure 8-10). The unwanted direct modifications of the dynamic library code segment instructions have been avoided by taking an extra step of indirection. In the lingo of programming languages, the approach can be described as using *pointer-to-pointer* instead of *pointer*.

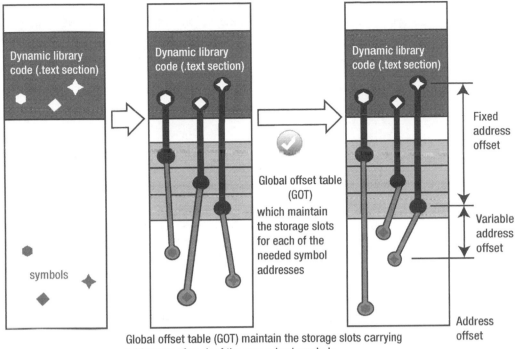

Global offset table (GOT) maintain the storage slots carrying addresses of each of the unresolved symbols.

The GOT offset is constant, and known at link time. Hence, the CPU instructions referencing GOT slot are independent of particular process memory map layout.

The content of storage slots, however, depend on the particular process memory map layout. Once everything is loaded and symbol addresses are known, the loader visit the GOT and updates the storage slots with the right values of symbol addresses.

Figure 8-10. PIC concept

Basically, the symbol addresses are supplied to the needy instructions in two steps. In order to get the symbol address, first a mov instruction accesses the address-of-address location and loads its contents (a needed symbol address) into an available CPU register. Immediately after, the retrieved symbol address now stored in a register may be used as the operand in the subsequent instructions (mov for data, call for function calls).

The special twist in the solution is that the symbol addresses are kept in a so-called global offset table (GOT) for which the linker reserves a dedicated .got section. The distance between the .text section and .got section is constant and known at link time. For each of the symbols that need to be resolved, the global offset table maintains a dedicated slot at the known and fixed offset from the table start.

Given the fixed GOT distance and fixed slot offset (both known at link time), it becomes possible for a compiler to implement the code instructions to reference the fixed locations. Most importantly, the implemented code does not depend on the actual symbol addresses, and can be used without any changes directly mapped into a plethora of other processes.

The final adjustment to the peculiarities of a specific memory map layout is completed by the loader. In this scheme, however, the loader does not irreversibly modify the code (.text section). Instead, once the symbol addresses are known, the loader patches the .got section, which is (much like the data sections) always implemented per process.

■ **Note**   In order to implement this scheme, a substantial design effort was needed, which is spread beyond the linker-loader boundaries. In fact, in order to implement the PIC concept, the story must start at the compiler level. In particular, the –fPIC flag must be passed to compiler. The "fPIC" or simply "PIC" mnemonic eventually became a synonym with dynamic linking.

Lazy Binding

The fact that the referencing of the symbols in the PIC approach passes through the extra level of indirection provides the potential for achieving the additional performance benefits at runtime. The strategy of implementing the extra performance kick is based on the fact that the loader does not waste precious time setting up the contents of the .got and .got.plt sections until the program starts.

The assembler instructions that reference the symbols are set to point to the intermediary point anyway, and there is nothing terribly wrong with the overall shape of the code that would put a stop to the program loading.

In fact, the loader normally does not even bother to complete setting up the contents of the .got and .got.plt sections until absolutely necessary. Such moments happen after the program already started, and only when the execution flow comes to the instructions referencing the symbols whose addresses are kept in the .got and .got.plt sections.

The obvious benefit of the loader's procrastination (commonly referred to as *lazy binding*) is that the loading process gets completed faster, which makes the application start quicker. A small one-time performance penalty happens when the loader quickly makes up for its initial (albeit premeditated) carelessness. That happens only as per need, and only once, at the very first occurrence of symbol referencing. The fewer of the dynamic library's symbols that actually get referenced at runtime, the more performance savings the loader is able to achieve.

The lazy binding concept is an extra feature of the PIC approach, which obviously adds another good reason for developers to choose the PIC over the LTR implementation. In fact, the PIC approach is a favorite implementation of the Scenario 1 type problem when the client binary is the executable file (i.e., application).

Rules and Limitations of the Recursive Chain of Dynamic Linking

The scenarios you've examined in detail so far pertain to the simplest possible dynamic linking scenarios. After taking a closer look at the atomic level, now let's step back a little and take a look at the molecular level of the dynamic linking, as it features certain rules and limitations which are not obvious at the atomic level of the story.

In reality, the structure of a typical program may be described as the recursive chain of dynamic linking, in which each of the dynamic libraries in the chain loads several other dynamic libraries. Visually, the recursive chain of loading can be represented as the elaborate tree structure with plenty of side connections between the branches. The length of individual branches in some cases may end up being really large. As interesting as the complexity of the recursive chain of dynamic linking may be, and as impressive the length of its branches may look, these are not the most potent details in the whole story.

Far more important is the emerging fact that in the chain of dynamic loading *each of the participating dynamic libraries may find itself playing the roles of both Scenario 1 and Scenario 2*. Figure 8-11 illustrates the point.

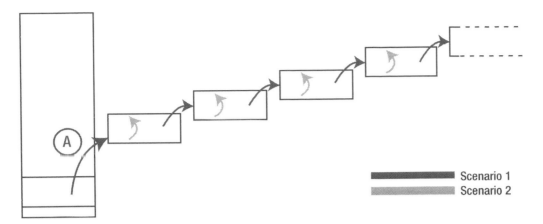

Figure 8-11. *A branch of typical recursive chain of dynamic linking*

In other words, a dynamic library in the chain of loading may need to both resolve the references of the library it loads as well as to re-resolve the reference of its own symbols. This makes the whole story a bit more interesting.

A certain set of strong implementation preferences residing at this molecular level stipulate the implementation details, which I will briefly review in the next section.

Strong Implementation Preferences

Regardless of the scenario, there are always two ways of how the linker-loader coordination may be implemented: either the LTR or the PIC approach may be applied. The choice of the linker-loader coordination technique is not absolutely free. In addition to the designer's choice based on each technique's pros and cons, there are a few other limitations that need to be explicitly pointed out:

- Position Independent Code (PIC) is the strongly preferred technique of executable resolving the references of the first level of loaded libraries (the scenario marked with the encircled letter *A* in Figure 8-11).

  In terms of choosing between LTR or PIC, the dynamic libraries in the chain of loading may feature a variety of combinations. A dynamic library implementing LTR may in turn dynamically load the next dynamic library, which implements the PIC, which in turn may dynamically load the library which implements ...you choose—whatever your choice is, it is allowed.

- A single dynamic library utilizes strictly one of the linker-loader coordination techniques to resolve both Scenario 1 as well as Scenario 2 (if needed). It cannot happen that the same dynamic library resolves Scenario 1 by the LRT approach and Scenario 2 problems by the PIC approach (or vice versa).

Figure 8-12 illustrates the described rules.

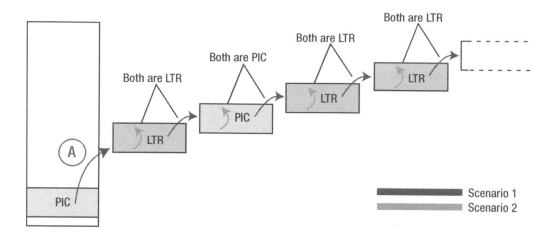

Figure 8-12. *Strong implementation preferences (at the molecular level) governing the implementation of the recursive chain of dynamic linking*

CHAPTER 9

■ ■ ■

Handling Duplicate Symbols When Linking In Dynamic Libraries

The concept of dynamic linking clearly represented a huge step forward in the domain of software design. The unprecedented flexibility it brought opened up a lot of avenues for technical progress and a lot of new doors to substantially new concepts.

By the same token, the complexity of how the dynamic libraries function internally has brought several distinct challenges to the domain of the software toolchain (compilers, linkers, loaders). The early recognized need for the linker and loader to cooperate more tightly and the techniques of implementing it were discussed in the previous chapter.

However, this is not where the story ends.

Another interesting paradigm closely associated with the domain of dynamic libraries is the issue of handling duplicate symbols. More specifically, when dynamic libraries are input ingredients in the linking process, the linker deviates from the usual, commonsensical approach typically followed in cases when individual object files and/or static libraries are combined into the binary file.

Duplicate Symbols Definition

The most common problem that may happen during the process of resolving the references is the appearance of duplicate symbols, which happens when, in the final stage of linking, the list of all available symbols contains two or more symbols of the same name.

As a side note, the linker algorithm for its own internal purposes typically applies modifications of the original symbol names. As a direct consequence, the reported duplicate issues printed out by the linker may refer to names somewhat differing from the original names. The symbol name modifications may range from simple name decorations (e.g., prepending the underscore) all the way to the systematic handling of C++ function affiliation issues. Fortunately, the modifications are typically performed in a strictly uniform and predictable fashion.

Typical Duplicate Symbols Scenarios

The causes of duplicate symbols may vary. In the simplest possible case, it just happened that the different designers chose the most obvious name for their modules classes, functions, structures (e.g., class `Timer`, function `getLength()` or variables `lastError` or `libVersion`). Trying to combine these designers' modules inevitably leads to discovering that duplicate symbols exist.

The other possibilities cover the typical cases when the data type instance (of a class, structure, or a simple data type) is defined in a header file. More than one inclusion of the header file inevitably creates the duplicate symbols scenario.

Duplicate C Symbols

The C language imposes fairly simple criteria for two or more symbols to be considered duplicates of each other. As long as the names of functions, structures, or data types are identical, the symbols are considered identical.

For example, buidling the following code will fail:

file: main.c
```
#include <stdio.h>

int function_with_duplicated_name(int x)
{
    printf("%s\n", __FUNCTION__);
    return 0;
}

int function_with_duplicated_name(int x, int y)
{
    printf("%s\n", __FUNCTION__);
    return 0;
}

int main(int argc, char* argv[])
{
    function_with_duplicated_name(1);
    function_with_duplicated_name(1,2);
    return 0;
}
```

It will produce the following error message:

```
main.c:9:5: error: conflicting types for 'function_with_duplicated_name'
main.c:3:5: note: previous definition of 'function_with_duplicated_name' was here
main.c: In function 'main':
main.c:17:5: error: too few arguments to function 'function_with_duplicated_name'
main.c:9:5: note: declared here
gcc: error: main.o: No such file or directory
gcc: fatal error: no input files
compilation terminated.
```

Duplicate C++ Symbols

Being an object-oriented programming language, C++ imposes more relaxed duplicate symbols criteria. In terms of namespaces, classes/structures, and simple data types, using identical names still remains as the sole criteria of duplicate symbols. However, in the domain of functions, the duplicate symbols criteria is no longer limited to the function names alone, but also take into account the list of arguments.

The principles of function (methods) overloading allows using the same name for different methods of the same class having different lists of input arguments, as long as the return value type does not differ.

The same principle applies to the corner cases of two or more functions belonging to the same namespace not being members of any class. Even though such functions are not affiliated with any class, the more elastic C++ duplicate criteria applies – they are considered duplicate only if their names are identical and their list of input arguments are identical.

Building the following code will be successfully completed:

file: main.cpp
```cpp
#include <iostream>
using namespace std;

class CTest
{
public:
    CTest(){ x = 0;};
    ~CTest(){},
public:
    int runTest(void){ return x;};
private:
    int x;
};

int function_with_duplicated_name(int x)
{
    cout << __FUNCTION__ << "(x)" << endl;
    return 0;
}

int function_with_duplicated_name(int x, int y)
{
    cout << __FUNCTION__ << "(x,y)" << endl;
    return 0;
}

int main(int argc, char* argv[]){
    CTest test;
    int x = test.runTest();

    function_with_duplicated_name(x);
    function_with_duplicated_name(x,1);
    return 0;
}
```

file: build.sh
```sh
g++ -Wall -g -O0 -c main.cpp
g++ main.o -o clientApp
```

Running the produced binary will create the following output:

```
function_with_duplicated_name(x)
function_with_duplicated_name(x,y)
```

However, trying to add the declaration of the following method to `main.cpp`

```
float function_with_duplicated_name(int x)
{
    cout << __FUNCTION__ << "(x)" << endl;
    return 0.0f;
}
```

will violate the basic rules of C++ function overloading, which will result with the following build failure:

```
main.cpp: In function 'float function_with_duplicated_name(int)':
main.cpp:23:42: error: new declaration 'float function_with_duplicated_name(int)'
main.cpp:17:5: error: ambiguates old declaration 'int function_with_duplicated_name(int)'
g++: error: main.o: No such file or directory
g++: fatal error: no input files
compilation terminated.
```

Duplicate Symbols Default Handling

When *individual object files or static libraries are being linked* together into the resultant binary file, the linker strictly follows the zero-tolerance policy against the duplicate symbols.

When the linker detects the duplicate symbols, it prints out an error message specifying the files/lines of code where the duplicate symbols occurrences happen, and the linking is declared a failure. This basically means that developers need to go back to the drawing board and try to resolve the problem, which very likely means that the code needs to be recompiled.

The following example illustrates what happens when you try to link into the same client binary the two static libraries featuring duplicate symbols. The project is comprised of two very simple static libraries featuring duplicate symbols as well as the client application that attempts to link them both:

Static Library libfirst.a:

file: staticlibfirstexports.h
```
#pragma once

int staticlibfirst_function(int x);
int staticlib_duplicate_function(int x);
```

file: staticlib.c
```
#include <stdio.h>

int staticlibfirst_function(int x)
{
    printf("%s\n", __FUNCTION__);
    return (x+1);
}

int staticlib_duplicate_function(int x)
{
    printf("%s\n", __FUNCTION__);
    return (x+2);
}
```

file: build.sh
```
gcc -Wall -g -O0 -c staticlib.c
ar -rcs libfirst.a staticlib.o
```

Static Library libsecond.a:

file: staticlibsecondexports.h
```
#pragma once

int staticlibsecond_function(int x);
int staticlib_duplicate_function(int x);
```

file: staticlib.c
```
#include <stdio.h>

int staticlibsecond_function(int x)
{
    printf("%s\n", __FUNCTION__);
    return (x+1);
}

int staticlib_duplicate_function(int x)
{
    printf("%s\n", __FUNCTION__);
    return (x+2);
}
```

file: build.sh
```
gcc -Wall -g -O0 -c staticlib.c
ar -rcs libsecond.a staticlib.o
```

ClientApplication:

file: main.c
```
#include <stdio.h>
#include "staticlibfirstexports.h"
#include "staticlibsecondexports.h"

int main(int argc, char* argv[])
{
    int nRetValue = 0;
    nRetValue += staticlibfirst_function(1);
    nRetValue += staticlibsecond_function(2);
    nRetValue += staticlib_duplicate_function(3);
    printf("nRetValue = %d\n", nRetValue);
    return nRetValue;
}
```

file: build.sh
```
gcc -Wall -g -O0 -I../libFirst -I../libSecond -c main.c
gcc main.o -L../libFirst -lfirst -L../libSecond -lsecond -o clientApp
```

Due to the presence of duplicate symbols in both static libraries, trying to build the client application results with the linker error:

```
/home/milan/Desktop/duplicateSymbolsHandlingResearch/01_duplicateSymbolsCriteria/02_duplicatesIn
TwoStaticLibs/01_plainAndSimple/libSecond/staticlib.c:10: multiple definition of
'staticlib_duplicate_function'
../libFirst/libfirst.a(staticlib.o):/home/milan/Desktop/duplicateSymbolsHandlingResearch/01_dupl
icateSymbolsCriteria/02_duplicatesInTwoStaticLibs/01_plainAndSimple/libFirst/staticlib.c:10: first
defined here
collect2: ld returned 1 exit status
```

Commenting out the call to the duplicate function does not help avoiding the linker failure. Obviously, the linker first tries to tile together everything coming from the input static libraries and individual object files (main.c). If the duplicate symbols happen this early in the linking game, the linker declares a failure, regardless of the fact that nobody tried to reference the duplicated symbols.

Duplicate Local Symbols Are Allowed

Interestingly, local functions declared with the static keyword in the C language meaning of that keyword (i.e., visibility scope limited only to the functions residing in the same source file) do not get registered as duplicates. Modify the source files of the static libraries in your example with the following code:

Static Library libfirst.a:
file: staticlib.c
```
static int local_staticlib_duplicate_function(int x)
{
    printf("libfirst: %s\n", __FUNCTION__);
    return 0;
}

int staticlibfirst_function(int x)
{
    printf("%s\n", __FUNCTION__);
    local_staticlib_duplicate_function(x);
    return (x+1);
}
```

Static Library libsecond.a:
file: staticlib.c
```
static int local_staticlib_duplicate_function(int x)
{
    printf("libsecond: %s\n", __FUNCTION__);
    return 0;
}

int staticlibsecond_function(int x)
{
    printf("%s\n", __FUNCTION__);
    local_staticlib_duplicate_function(x);
    return (x+1);
}
```

ClientApplication:

file: main.c

```
#include <stdio.h>
#include "staticlibfirstexports.h"
#include "staticlibsecondexports.h"

int main(int argc, char* argv[])
{
    staticlibfirst_function(1);
    staticlibsecond_function(2);
    return 0;
}
```

The client application will now build successfully and produce the following output:

```
staticlibfirst_function
libfirst: local_staticlib_duplicate_function
staticlibsecond_function
libsecond: local_staticlib_duplicate_function
```

Obviously, the linker keeps the separate compartments for the local functions. Even though their symbol names are completely identical, the collision does not happen.

Duplicate Symbols Handling When Linking in Dynamic Libraries

When dynamic libraries are added to the mix of input ingredients at the linking stage, the way the linker handles the duplicate symbols becomes a lot more interesting and lot more involved. First and foremost, the linker abandons the policy of zero-tolerance for duplicate symbols, and does not immediately declare the linking failure. Instead, it applies the approximate, less-than-ideal approach to resolve the symbols naming collision.

In order to illustrate the linker's altogether different approach to this specific scenario, the simple demo project is created. It is comprised of the two dynamic libraries featuring the duplicate symbols and the client application that links them both:

Shared Library libfirst.so:

file: shlibfirstexports.h
```
#pragma once

int shlibfirst_function(int x);
int shlib_duplicate_function(int x);
```

file: shlib.c
```
#include <stdio.h>

static int local_shlib_duplicate_function(int x)
{
    printf("shlibFirst: %s\n", __FUNCTION__);
    return 0;
}
```

```c
int shlibfirst_function(int x)
{
    printf("shlibFirst: %s\n", __FUNCTION__);
    local_shlib_duplicate_function(x);
    return (x+1);
}

int shlib_duplicate_function(int x)
{
    printf("shlibFirst: %s\n", __FUNCTION__);
    local_shlib_duplicate_function(x);
    return (x+2);
}
```

file: build.sh
```
gcc -Wall -g -O0 -fPIC -c shlib.c
gcc -shared shlib.o -Wl,-soname,libfirst.so.1 -o libfirst.so.1.0.0
ldconfig -n .
ln -s libfirst.so.1 libfirst.so
```

Shared Library libsecond.so:

file: shlibsecondexports.h
```c
#pragma once

int shlibsecond_function(int x);
int shlib_duplicate_function(int x);
```

file: shlib.c
```c
#include <stdio.h>

static int local_shlib_duplicate_function (int x)
{
    printf("shlibSecond: %s\n", __FUNCTION__);
    return 0;
}

int shlibsecond_function(int x)
{
    printf("shlibSecond: %s\n", __FUNCTION__);
    local_shlib_duplicate_function(x);
    return (x+1);
}

int shlib_duplicate_function(int x)
{
    printf("shlibSecond: %s\n", __FUNCTION__);
    local_shlib_duplicate_function(x);
    return (x+2);
}
```

file: build.sh
```
gcc -Wall -g -O0 -fPIC -c shlib.c
gcc -shared shlib.o -Wl,-soname,libsecond.so.1 -o libsecond.so.1.0.0
ldconfig -n .
ln -s libsecond.so.1 libsecond.so
```

ClientApplication:

file: main.c
```
#include <stdio.h>
#include "shlibfirstexports.h"
#include "shlibsecondexports.h"

int main(int argc, char* argv[])
{
    int nRetValue = 0;
    nRetValue += shlibfirst_function(1);
    nRetValue += shlibsecond_function(2);
    nRetValue += shlib_duplicate_function(3);
    return nRetValue;
}
```

file: build.sh

```
gcc -Wall -g -O0 -I../libFirst -I../libSecond -c main.c
gcc main.o -Wl,-L../libFirst -Wl,-lfirst    \
          -Wl,-L../libSecond -Wl,-lsecond \
          -Wl,-R../libFirst               \
          -Wl,-R../libSecond              \
          -o clientApp
```

Even though the two shared libraries feature the duplicates and even one of the duplicates (shlib_duplicate_function) is not local function, building the client application completes successfully.

Running the client application, however, brings a bit of surprise:

shlibFirst: shlibfirst_function
shlibFirst: local_shlib_duplicate_function
shlibSecond: shlibsecond_function
shlibSecond: local_shlib_duplicate_function
shlibFirst: shlib_duplicate_function
shlibFirst: local_shlib_duplicate_function

Obviously, the linker found some way of resolving the duplicate symbols. It solved it by picking one of the symbol occurrences (the one in shlibfirst.so) and directed all the references to shlib_duplicate_function to that particular symbol occurrence.

This linker's decision is clearly a very controversial step. In real-world scenarios, the identically named functions of different dynamic libraries may carry substantially different functionality. Imagine, for example, that each of the dynamic libraries libcryptography.so, libnetworkaccess.so, and libaudioport.so feature the Initialize() method. Imagine now that the linker decided that the call to Initialize() always means only initializing one of the libraries (and never initializing the other two).

Obviously, these kinds of scenarios should be carefully avoided. In order to do it right, the way of how the linker "thinks" should be thoroughly understood first.

The details of the linker's internal algorithm for handling dynamic library duplicate symbols will be discussed later in this chapter.

General Strategies of Eliminating Duplicate Symbols Problems

In general, the best approach to resolving the duplicate symbols is to reinforce the symbol affiliations to their particular modules, as it usually eliminates the vast majority of potential duplicate symbols problems.

In particular, resorting to the use of *namespaces* is the most recommended technique, as it has been proven to work across plenty of different scenarios, regardless of the form in which the code is made available to the software community (static library vs. shared library). This feature is confined to the domain of the C++ language and requires the use of the C++ compiler.

Alternatively, if for whatever reason the use of a strictly C compiler is strongly preferred, prepending the function names with the unique prefix may be used as a viable, yet somewhat less powerful and less flexible, alternative.

Duplicate Symbols and Dynamic Linking Modes

Before going into the details of the linker's new approach to handling the duplicate symbols, it is important to point out a few significant facts.

The runtime dynamic loading of dynamic libraries (through the `dlopen()` or `LoadLibrary()` calls) imposes practically no risk of having duplicate symbols. The retrieved dynamic library symbols are typically assigned (through the `dlsym()` or `GetProcAddress()` calls) to the variable whose name is very likely already chosen to not duplicate any of the existing symbols in the client binary.

On the contrary, it is the *statically aware linking* of dynamic libraries that represents the typical scenario in which the duplicate symbols occurrences happen.

The genuine reason for deciding to link in a dynamic library is the interest in the set of the dynamic library's ABI symbols or its subset. Very frequently, however, the dynamic library may carry a lot more symbols of remote or no importance to the client binary project, and the unawareness of their presence may lead to an unintentional choice of a duplicate named function or data coming from different dynamic libraries.

There is only so much precaution that dynamic library developers can take in order to make the things better. Reducing the export of the dynamic library symbols to only the essential set of symbols is definitely a measure that may significantly reduce the probability of symbol names collision. However, this highly recommended design practice does not directly act against the very root of the problem. Regardless of how frugal you are in exporting your dynamic library symbols, it is still possible that different developers choose the most straightforward names for the symbols, which results with two or more binary files contending about the right to use the symbol name.

Finally, it is important to point out that you are not dealing with the peculiarity of a specific linker on specific platform; the Windows linker (Visual Studio 2010 certainly) almost completely follows the same set of rules in determining how to handle the duplicate symbols encountered during the process of dynamic linking.

Linker's Criteria in the Approximate Algorithm of Resolving Dynamic Libraries' Duplicate Symbols

In the search for the best candidate to represent the duplicated symbol name, the linker makes the decision based on the following circumstances:

- *Location of the duplicated symbols*: The linker assigns different levels of importance to the symbols located in different parts of the process memory map. A more detailed explanation follows immediately.

- *Specified linking order of the dynamic libraries*: If two or more symbols reside in the code parts of equal priorities, the symbol residing in the dynamic library that was passed to the linker earlier in the list of specified dynamic libraries will win the bout to represent the duplicated symbol over the symbol residing in the dynamic library declared later on in the list.

Location, Location, Location: Code Priority Zoning Rules

The variety of linker symbols participating in building the client binary may reside in a variety of locations. The first criterion that the linker applies toward resolving name collisions between symbols is based on the comparison between the following symbols priority scheme.

First Level Priority Symbols: Client Binary Symbols

The initial ingredient of building the binary file is the collection of its object files, which are either indigenous to the project or come in the form of the static library. In the case of Linux, the sections coming from these ingredients typically occupy the lower part of the process memory map.

Second Level Priority Symbols: Dynamic Library Visible Symbols

The dynamic library exported symbols (residing in the dynamic section of the dynamic libraries) are taken by the linker as the next priority level in the priority scheme.

Third Level Priority (Unprioritized, Noncompeting) Symbols

The symbols declared as static are typically never the subject of the duplicated symbol name conflicts, regardless of whether they belong to the client binary or to the statically aware linked dynamic library.

To the same group belong the dynamic library's stripped off symbols, which obviously do not participate in the stage of linking the client binary. Figure 9-1 illustrates the symbols priority zoning approach.

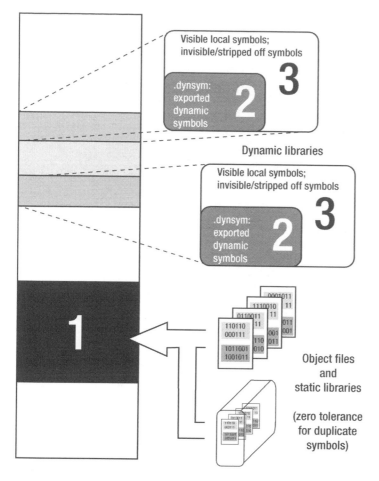

Figure 9-1. *Linker's priority zoning*

Analyses of Specific Duplicate Names Cases

The following sections cover several use cases.

Case 1: Client Binary Symbol Collides with Dynamic Library ABI Function

This scenario can be basically described as the symbol belonging to priority zone 1 colliding against the symbol belonging to priority zone 2 (Figure 9-2).

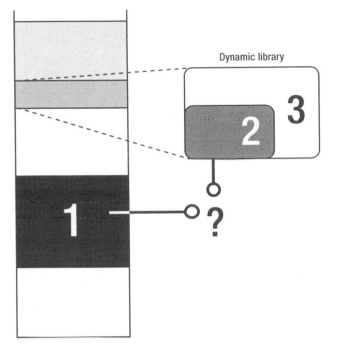

Figure 9-2. *Case 1: The client binary symbol collides with the dynamic library ABI symbol*

As a general rule, the symbol related to the higher priority code zone always wins; in other words, it gets chosen by the linker as the target of all the references to the duplicate named symbol.

The following project is created to demonstrate this particular scenario. It consists of a static library, a dynamic library, and the client application that links them both (the dynamic library is statically aware linked). The libraries feature a duplicate name symbol:

Static Library libstaticlib.a:

file: staticlibexports.h
```
#pragma once

int staticlib_first_function(int x);
int staticlib_second_function(int x);

int shared_static_duplicate_function(int x);
```

file: staticlib.c
```
#include <stdio.h>
#include "staticlibexports.h"

int staticlib_first_function(int x)
{
    printf("%s\n", __FUNCTION__);
    return (x+1);
}
```

```
int staticlib_second_function(int x)
{
    printf("%s\n", __FUNCTION__);
    return (x+2);
}

int shared_static_duplicate_function(int x)
{
    printf("staticlib: %s\n", __FUNCTION__);
    return 0;
}
```

file: build.sh
```
gcc -Wall -g -O0 -c staticlib.c
ar -rcs libstaticlib.a staticlib.o
```

Shared Library libshlib.so:

file: shlibexports.h
```
#pragma once

int shlib_function(void);
int shared_static_duplicate_function(int x);
```

file: shlib.c
```
#include <stdio.h>
#include "staticlibexports.h"

int shlib_function(void)
{
        printf("sharedLib: %s\n", __FUNCTION__);
    return 0;
}

int shared_static_duplicate_function(int x)
{
    printf("sharedLib: %s\n", __FUNCTION__);
    return 0;
}
```

file: build.sh
```
gcc -Wall -g -O0 -I../staticLib -c shlib.c
gcc -shared shlib.o -Wl,-soname,libshlib.so.1 -o libshlib.so.1.0.0
ldconfig -n .
ln -s libshlib.so.1 libshlib.so
```

CHAPTER 9 ▓ HANDLING DUPLICATE SYMBOLS WHEN LINKING IN DYNAMIC LIBRARIES

ClientApplication:

file: main.c
```
#include <stdio.h>
#include "staticlibexports.h"
#include "shlibexports.h"
int main(int argc, char* argv[])
{
    int nRetValue = 0;
    nRetValue += staticlib_first_function(1);
    nRetValue += staticlib_second_function(2);

    shlib_function();
    shared_static_duplicate_function(1);
    printf("nRetValue = %d\n", nRetValue);
    return nRetValue;
}
```

file: build.sh
```
gcc -Wall -g -OO -I../staticLib -I../sharedLib -c main.c
gcc main.o -Wl,-L../staticLib -lstaticlib \
         -Wl,-L../sharedLib -lshlib      \
         -Wl,-R../sharcdLib              \
         -o clientApp
```

The client application builds successfully and produces the following output:

```
staticlib_first_function
staticlib_second_function
sharedLib: shlib_function
```
staticlib: shared_static_duplicate_function
```
nRetValue = 6
```

Obviously, the linker picks the static library symbol as it belongs to the higher priority code zone.
Change the build order, as shown here:

file: buildDifferentLinkingOrder.sh
```
gcc -Wall -g -OO -I../staticLib -I../sharedLib -c main.c
gcc main.o -Wl,-L../sharedLib -lshlib      \
         -Wl,-L../staticLib -lstaticlib \
         -Wl,-R../sharedLib             \
         -o clientAppDifferentLinkingOrder
```

Note that the change in code does not change the final outcome:

```
$ ./clientAppDifferentLinkingOrder
staticlib_first_function
staticlib_second_function
sharedLib: shlib_function
```
staticlib: shared_static_duplicate_function
```
nRetValue = 6
```

Windows-Specific Twist

The Visual Studio linker has a slightly different way of implemented this ruling in this particular case (i.e., when the static library features the symbol of the same name with the dynamic library ABI symbol).

When the static library appears as first on the list of libraries, the DLL's symbols is silently ignored, which is exactly as expected.

However, if the DLL is specified as the first on the list of libraries, what happens is not what you might expect (i.e., that the static library symbol always prevails). Instead, the linking fails with a message saying something like

```
StaticLib (staticlib.obj): error LNK2005: function_xyz already defined \
        in SharedLib.lib (SharedLib.dll)
ClientApp.exe: fatal error LNK1169: one or more multiply defined symbols found
BUILD FAILED.
```

Case 2: ABI Symbols of Different Dynamic Libraries Collide

This scenario can be basically described as two symbols both belonging to the priority zone 2 colliding against each other (Figure 9-3).

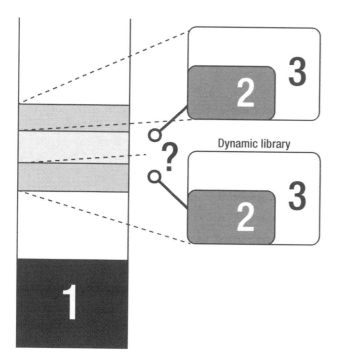

Figure 9-3. *Case 2: ABI symbols of different dynamic libraries collide*

Clearly, since none of the symbols have the zoning advantage, the decisive factor in this case will be the linking order.

In order to demonstrate this particular scenario, the following demo project is created, which consists of two shared libraries featuring the duplicate ABI symbols and the client app that statically aware links both dynamic libraries. To provide a few more important details, one of the shared library ABI functions internally calls the duplicated ABI function:

Shared Library libfirst.so:

file: shlibfirstexports.h
```
#pragma once

int shlib_function(void); // duplicate ABI function
int shlibfirst_function(void);
```

file: shlib.c
```
#include <stdio.h>

int shlib_function(void)
{
        printf("shlibFirst: %s\n", __FUNCTION__);
    return 0;
}

int shlibfirst_function(void)
{
        printf("%s\n", __FUNCTION__);
        return 0;
}
```

file: build.sh
```
gcc -Wall -g -O0 -c shlib.c
gcc -shared shlib.o -Wl,-soname,libfirst.so.1 -o libfirst.so.1.0.0
ldconfig -n .
ln -s libfirst.so.1 libfirst.so
```

Shared Library libsecond.so:

file: shlibsecondexports.h
```
#pragma once

int shlib_function(void);
int shlibsecond_function(void);
int shlibsecond_another_function(void);
```

file: shlib.c
```
#include <stdio.h>

int shlib_function(void)
{
    printf("shlibSecond: %s\n", __FUNCTION__);
    return 0;
}
```

171

```
int shlibsecond_function(void)
{
    printf("%s\n", __FUNCTION__);
    return 0;
}

int shlibsecond_another_function(void)
{
    printf("%s\n", __FUNCTION__);
    shlib_function(); // internal call to the duplicate ABI function
    return 0;
}
```

file: build.sh
```
gcc -Wall -g -O0 -fPIC -c shlib.c
gcc -shared shlib.o -Wl,-soname,libsecond.so.1 -o libsecond.so.1.0.0
ldconfig -n .
ln -s libsecond.so.1 libsecond.so
```

ClientApplication:

file: main.c
```
#include <stdio.h>
#include "shlibfirstexports.h"
#include "shlibsecondexports.h"

int main(int argc, char* argv[])
{
    shlib_function();      // duplicate ABI function
    shlibfirst_function();
    shlibsecond_function();
    shlibsecond_another_function(); // this one internally calls shlib_function()
    return 0;
}
```

file: build.sh
```
gcc -Wall -g -O0 -I../libFirst -I../libSecond -c main.c
gcc main.o -Wl,-L../libFirst -Wl,-lfirst    \
        -Wl,-L../libSecond -Wl,-lsecond \
        -Wl,-R../libFirst                    \
        -Wl,-R../libSecond                   \
        -o clientApp
```

Even though the two shared libraries feature the duplicates and even one of the duplicates (shlib_duplicate_function) is not local function, building the client application completes successfully.

Running the client application results in the following output:

```
$ ./clientApp
shlibFirst: shlib_function
shlibfirst_function
shlibsecond_function
shlibsecond_another_function
shlibFirst: shlib_function
```

Obviously, the linker picked the shlibFirst's version of the duplicate symbol to uniquely represent the duplicated symbol name. Furthermore, even though shlibsecond_another_function() internally calls the duplicated shlib_function(), it does not affect the final outcome of the linking stage.

Being the ABI symbol (the part of the .dynsym section), the duplicated function symbol always gets resolved in the same way, regardless of the fact that it resides in the same source file with the remaining ABI functions.

No Impact of Different Function Calls Order

As the part of the investigation, the impact of the reversed function call order is examined (see Listing 9-1).

Listing 9-1. main_differentOrderOfCalls.c

```
#include <stdio.h>
#include "shlibfirstexports.h"
#include "shlibsecondexports.h"

int main(int argc, char* argv[])
{
    // Reverse order of calls - first shlibsecond methods
    // get called, followed by the shlibfirst methods
    shlibsecond_function();
    shlibsecond_another_function();
    shlib_function();     // duplicate ABI function
    shlibfirst_function();
    return 0;
}
```

This particular change did not affect the final outcome in any way. Obviously, the important moments of the linking stage that critically impact the duplicate symbol resolution process happen during an earlier stage of linking.

Impact of Different Linking Order

Building the application with the different linking order, however, yields different results:

```
file: buildDifferentLinkingOrder.sh
gcc -Wall -g -O0 -I../shlibFirst -I../shlibSecond -c main.c
gcc main.o -Wl,-L../shlibSecond -lsecond \
           -Wl,-L../shlibFirst  -lfirst \
           -Wl,-R../shlibFirst           \
           -Wl,-R../shlibSecond          \
           -o clientAppDifferentLinkingOrder


$ ./clientAppDifferentLinkingOrder
shlibSecond: shlib_function
shlibfirst_function
shlibsecond_function
shlibsecond_another_function
shlibSecond: shlib_function
```

Obviously, the specified reversed linking order affected the linker's decision. The shlibSecond's version of duplicated shlib_function is now chosen to represent the duplicated symbol.

Case 3: Dynamic Library ABI Symbol Collides with Another Dynamic Library Local Symbol

This scenario can be basically described as a symbol belonging to priority zone 2 colliding against a symbol belonging to priority zone 3 (Figure 9-4).

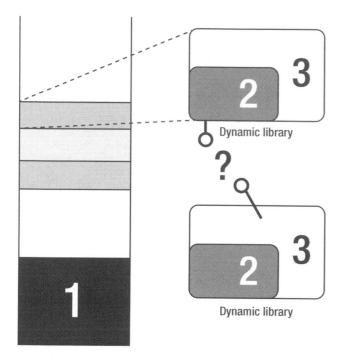

Figure 9-4. *Case 3: Dynamic library ABI symbol collides with another dynamic library local symbol*

As a general rule, much like in the Case 1 example, the symbol related to the higher priority code zone always wins the bout; in other words, it gets chosen by the linker as the target of all the references to the duplicate named symbol.

In order to illustrate this particular scenario, the following demo project is created; it consists of two shared libraries (featuring the duplicate symbols) and the client application that statically aware links both libraries:

Shared Library libfirst.so:

file: shlibfirstexports.h
```
#pragma once

int shlib_function(void);
int shlibfirst_function(void);
```

file: shlib.c
```
#include <stdio.h>
```

```c
int shlib_function(void)
{
    printf("shlibFirst: %s\n", __FUNCTION__);
    return 0;
}

int shlibfirst_function(void)
{
    printf("%s\n", __FUNCTION__);
    return 0;
}
```

file: build.sh
```sh
gcc -Wall -g -O0 -c shlib.c
gcc -shared shlib.o -Wl,-soname,libfirst.so.1 -o libfirst.so.1.0.0
ldconfig -n .
ln -s libfirst.so.1 libfirst.so
```

Shared Library libsecond.so:

file: shlibsecondexports.h
```c
#pragma once

int shlibsecond_function(void);
```

file: shlib.c
```c
#include <stdio.h>

static int shlib_function(void)
{
    printf("shlibSecond: %s\n", __FUNCTION__);
    return 0;
}

int shlibsecond_function(void)
{
    printf("%s\n", __FUNCTION__);
    shlib_function();
    return 0;
}
```

file: build.sh
```sh
gcc -Wall -g -O0 -c shlib.c
gcc -shared shlib.o -Wl,-soname,libsecond.so.1 -o libsecond.so.1.0.0
ldconfig -n .
ln -s libsecond.so.1 libsecond.so
```

ClientApplication:

file: `main.c`
```c
#include <stdio.h>
#include "shlibfirstexports.h"
#include "shlibsecondexports.h"

int main(int argc, char* argv[])
{
    shlibfirst_function();
    shlibsecond_function();
    return 0;
}
```

file: `build.sh`
```
gcc -Wall -g -O0 -I../shlibFirst -I../shlibSecond -c main.c
gcc main.o -Wl,-L../shlibFirst -lfirst   \
           -Wl,-L../shlibSecond -lsecond \
           -Wl,-R../shlibFirst           \
           -Wl,-R../shlibSecond          \
           -o clientApp
```

Building the client application completed successfully. Running the client application results with the following output:

```
$ ./clientApp
shlibFirst: shlib_function
shlibsecond_function
shlibSecond: shlib_function
```

Here we have a bit of interesting situation.

First, when the client binary invokes the duplicate named `shlib_function`, the linker has no doubts that this symbol should be represented by the `shlibFirst` library method, simply because it resides in the code zone of higher priority. The first line of the client app output witnesses to this fact.

However, much before the linker deliberation happened, during the building of the dynamic library itself, the internal calls of `shlibsecond_function()` to its local `shlib_function()` was already resolved, simply because the two symbols are local to each other. This is the reason why the internal call of one `shlibSecond` function to another `shlibSecond` function does not get affected by the process of building the client binary.

As expected, when the linker's decision is determined by the differences in the code zone priorities, reversing the linking order has no impact on the final outcome.

Case 4: Dynamic Library Non-exported Symbol Collides with Another Dynamic Library Non-exported Symbol

This scenario can be basically described as two symbols both belonging to priority zone 3 collide against each other (Figure 9-5).

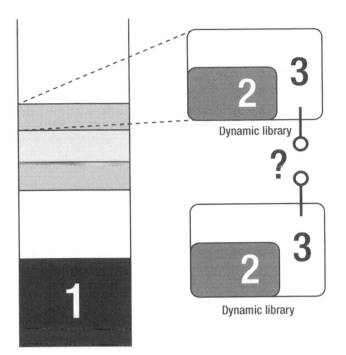

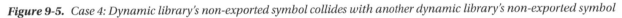

Figure 9-5. *Case 4: Dynamic library's non-exported symbol collides with another dynamic library's non-exported symbol*

The symbols belonging to code zone 3 are mostly invisible to the process of building the client binary. These symbols are either declared to be of local scope (and completely not interesting to the linker), or stripped off (invisible to the linker).

Even though the symbol names may be duplicated, these symbols do not end up in the linker's list of symbols and do not cause any conflicts. Their importance is strictly confined to the domain of the dynamic libraries of which they are a part.

In order to illustrate this particular scenario, the following demo project is created; it consists of one static library, one shared library, and the client application that links both libraries. The dynamic library is linked statically aware.

Each of the binaries features local functions whose names are the duplicates of the names of local functions found in the remaining modules. Additionally, the client application has the local function identically named as the function of the shared library whose symbols is intentionally stripped off.

Static Library libstaticlib.a:

file: staticlibexports.h
```
#pragma once

int staticlib_function(int x);
```

file: staticlib.c
```
#include <stdio.h>
#include "staticlibexports.h"
```

```
static int local_function(int x)
{
    printf("staticLib: %s\n", __FUNCTION__);
    return 0;
}

int staticlib_function(int x)
{
    printf("%s\n", __FUNCTION__);
    local_function(x);
    return (x+1);
}
```

file: build.sh
```
gcc -Wall -g -O0 -c staticlib.c
ar -rcs libstaticlib.a staticlib.o
```

Shared Library libshlib.so:
file: shlibexports.h
```
#pragma once

int shlib_function(void);
```

file: shlib.c
```
#include <stdio.h>
#include "staticlibexports.h"

static int local_function(int x)
{
    printf("sharedLib: %s\n", __FUNCTION__);
    return 0;
}

static int local_function_strippedoff(int x)
{
    printf("sharedLib: %s\n", __FUNCTION__);
    return 0;
}

int shlib_function(void)
{
    printf("sharedLib: %s\n", __FUNCTION__);
    local_function(1);
    local_function_strippedoff(1);
    return 0;
}
```

file: build.sh
```
gcc -Wall -g -O0 -I../staticLib -c shlib.c
gcc -shared shlib.o -Wl,-soname,libshlib.so.1 -o libshlib.so.1.0.0
strip -N local_function_strippedoff libshlib.so.1.0.0
ldconfig -n .
ln -s libshlib.so.1 libshlib.so
```

Client Application:
file: main.c
```c
#include <stdio.h>
#include "staticlibexports.h"
#include "shlibexports.h"

static int local_function(int x)
{
    printf("clientApp: %s\n", __FUNCTION__);
    return 0;
}

static int local_function_strippedoff(int x)
{
    printf("clientApp: %s\n", __FUNCTION__);
    return 0;
}

int main(int argc, char* argv[])
{
    shlib_function();
    staticlib_function(1);
    local_function(1);
    local_function_strippedoff(1);
    return 0;
}
```

file: build.sh
```sh
gcc -Wall -g -O0 -I../staticLib -I../sharcdLib  c main.c
gcc main.o -Wl,-L../staticLib -lstaticlib \
          -Wl,-L../sharedLib -lshlib      \
          -Wl,-R../sharedLib              \
          -o clientApp
```

As expected, the client application is built successfully and produced the following output:

```
sharedLib: shlib_function
sharedLib: local_function
sharedLib: local_function_strippedoff
staticlib_function
staticLib: local_function
clientApp: local_function
clientApp: local_function_strippedoff
```

Obviously, the linker did not perceive any duplicate symbol issue. All the local/stripped off symbols have been resolved within their particular modules, and did not conflict with any of identically-named local/stripped off symbols in the other modules.

Interesting Scenario: Singleton in Static Library

Now that you know how the linker handles the unprioritized/non-competing domain of dynamic libraries' local/stripped off symbols, it is much easier to understand the interesting scenario described in Chapter 6 as a "multiple instances of a singleton class" problem (one of the counter-indicated scenarios for using the static libraries).

Imagine for a moment the following real-world scenario: say you need to design a unique process-wide logging utility class. It should exist in one instance and should be visible to all of the different functional modules.

The implementation paradigm would be normally based on the singleton design pattern. Let's assume for a moment that the home of your singleton class is a dedicated static library.

In order to get access to the logging utility, several dynamic libraries hosting your functional modules link in that particular static library. Being merely part of the dynamic library inner functionality (i.e., not part of dynamic library's ABI interface), the singleton class symbols have *not* been exported. The singleton class symbols automatically start belonging to the unprioritized/noncompeting code zone.

Once the process starts and all the dynamic libraries are loaded, you end up having a situation in which several dynamic libraries live in the same process, each of them having a singleton class in their own "private backyards." Lo and behold, due to the non-competing nature of the dynamic libraries' local symbols zone, all of a sudden you end up having multiple (nicely coexisting) instances of your singleton logging utility class.

The only problem is that you wanted a single, unique singleton class instance, not many of them!!!

In order to illustrate this particular scenario, the next demo project is created with the following components:

- A static library hosting the singleton class

- Two shared libraries, each linking in the static library. Each of the shared libraries export only one symbol: a function that internally calls the singleton object methods. Singleton class symbols coming in from the linked static library are not exported.

- A client application that links in the static library in order to access the singleton class itself. It also statically aware links in both of the shared libraries.

The client app and both shared libraries make their own calls to the singleton class. As you will see shortly, the app will feature three different instances of the singleton class:

Static Library libsingleton.a:

file: singleton.h

```
#pragma once

class Singleton
{
public:
    static Singleton& GetInstance(void);

public:
    ~Singleton(){};
    int DoSomething(void);

private:
    Singleton(){};
    Singleton(Singleton const &);      // purposefully not implemented
    void operator=(Singleton const&); // purposefully not implemented

private:
    static Singleton* m_pInstance;
};
```

file: singleton.cpp
```cpp
#include <iostream>
#include "singleton.h"
using namespace std;

Singleton* Singleton::m_pInstance = NULL;

Singleton& Singleton::GetInstance(void)
{
    if(NULL == m_pInstance)
        m_pInstance = new Singleton();
    return *m_pInstance;
}

int Singleton::DoSomething(void)
{
    cout << "singleton instance address = " << this << endl;
    return 0;
}
```

file: build.sh
```sh
# for 64-bit OS must also pass -mcmodel=large compiler flag
g++ -Wall -g -O0 -c singleton.cpp
ar -rcs libsingleton.a singleton.o
```

Shared Library libfirst.so:

file: shlibfirstexports.h
```c
#pragma once

#ifdef __cplusplus
extern "C"
{
#endif // __cplusplus

int shlibfirst_function(void);

#ifdef __cplusplus
}
#endif // __cplusplus
```

file: shlib.c
```c
#include <iostream>
#include "singleton.h"
using namespace std;

#ifdef __cplusplus
extern "C"
{
#endif // __cplusplus
```

```cpp
int shlibfirst_function(void)
{
    cout << __FUNCTION__ << ":" << endl;
    Singleton& singleton = Singleton::GetInstance();
    singleton.DoSomething();
    return 0;
}

#ifdef __cplusplus
}
#endif // __cplusplus
```

file: build.sh
```bash
rm -rf *.o lib*
g++ -Wall -g -O0 -fPIC -I../staticLib -c shlib.cpp
g++ -shared shlib.o -L../staticLib -lsingleton    \
    -Wl,--version-script=versionScript            \
    -Wl,-soname,libfirst.so.1 -o libfirst.so.1.0.0
ldconfig -n .
ln -s libfirst.so.1 libfirst.so
```

file: versionScript
```
{
    global:
        shlibfirst_function;
    local:
        *;
};
```
Shared Library libsecond.so:

file: shlibfirstexports.h
```c
#pragma once

#ifdef __cplusplus
extern "C"
{
#endif // __cplusplus

int shlibsecond_function(void);

#ifdef __cplusplus
}
#endif // __cplusplus
```

file: shlib.c
```c
#include <iostream>
#include "singleton.h"
using namespace std;
```

```
#ifdef __cplusplus
extern "C"
{
#endif // __cplusplus

int shlibsecond_function(void)
{
    cout << __FUNCTION__ << ":" << endl;
    Singleton& singleton = Singleton::GetInstance();
    singleton.DoSomething();
    return 0;
}

#ifdef __cplusplus
}
#endif // __cplusplus
```

file: build.sh
```
rm -rf *.o lib*
g++ -Wall -g -O0 -fPIC -I../shlibFirst -I../staticLib -c shlib.cpp
g++ -shared shlib.o -L../staticLib -lsingleton         \
    -Wl,--version-script=versionScript                 \
    -Wl,-soname,libsecond.so.1 -o libsecond.so.1.0.0
ldconfig -n .
ln -s libsecond.so.1 libsecond.so
```

file: versionScript
```
{
    global:
        shlibsecond_function;
    local:
        *;
};
```

ClientApplication:

file: main.c
```
#include <iostream>
#include "shlibfirstexports.h"
#include "shlibsecondexports.h"
#include "singleton.h"

using namespace std;

int main(int argc, char* argv[])
{
    shlibfirst_function();
    shlibsecond_function();
    cout << "Accesing singleton directly from the client app" << endl;
    Singleton& singleton = Singleton::GetInstance();
    singleton.DoSomething();
    return 0;
}
```

file: build.sh

```
g++ -Wall -g -O0 -I../staticLib -I../shlibFirst -I../shlibSecond -c main.cpp
g++ main.o -L../staticLib -lsingleton \
          -L../shlibFirst -lfirst    \
          -L../shlibSecond -lsecond  \
          -Wl,-R../shlibFirst        \
          -Wl,-R../shlibSecond       \
          -o clientApp
```

The client application produced the following output:

```
shlibfirst_function:
singleton instance address = 0x9a01008
shlibsecond_function:
singleton instance address = 0x9a01018
Accesing singleton directly from the client app
singleton instance address = 0x9a01028
```

■ **Note**   It is left to the diligent reader to find out that the runtime dynamic loading (dlopen) would not change anything in this regard.

As a final note on this topic, a thread-safer singleton version in which the singleton instance would be a function static variable instead of a class-static variable was tried:

```
Singleton& Singleton::GetInstance(void)
{
    Static Singleton uniqueInstance;
    return uniqueInstance;
}
```

This approach resulted with only somewhat better results, in which both of the shared libraries print out the identical singleton instance address value, albeit the client app printed out substantially different singleton instance address values.

Solving the Problem

To not be completely pessimistic, there are several ways to solve this class of problem.

One of the possibilities is based on relaxing the symbol export criteria for a little bit by allowing the dynamic libraries to additionally export the singleton class symbols. After being exported, the singleton symbols will no longer belong to the unprioritized/non-competing category of symbols allowed to exist in gazillion of instances. Instead, they will be promoted into the "competing ABI symbols" category. According to the elaborated rules, the linker would then pick just one of the symbols and direct all the references to that particular singleton class symbol.

The ultimate solution to the problem would be to host the singleton class in a dynamic library. That way, the vast majority of the possible unwanted scenarios would be completely eliminated. None of the ABI design rules would be violated, and the design of new modules will not be facing the ludicrous extra design requirements.

Final Remark: Linking Does Not Provide Any Kind of Namespace Inheritance

The use of namespaces is definitely the most powerful tool to completely avoid the unpleasant surprises coming from too much reliance on the linker's internal reasoning in handling the duplicate symbols.

Regardless of the fact that one shared library may link another shared library, which may link yet another shared library, which eventually may link the static library, protecting the uniqueness of the symbols carried by a library sitting somewhere in the middle of the chain of linking requires that exactly that particular library's code be encapsulated in its own proprietary namespace.

Expecting that the namespace of the topmost library will shield the uniqueness of the symbols of the library in between possible clashes with the other dynamic libraries is just plain wrong.

The only solid plan, the one that really works, is that each and every library, static or dynamic, should feature its own dedicated namespace.

■ ■ ■

## CHAPTER 10

Dynamic Libraries Versioning

Most of the time, code development is work in progress. As a result of striving to provide more and more features, as well as to solidify the existing body of code, the code inevitably changes. More frequently than not, the design's quantum leaps tend to break the compatibility between the software components. The ideal of achieving backwards compatibility typically requires a dedicated and focused effort. A very important role in these efforts belongs to the versioning concept.

Given the fact that dynamic libraries provide functionality that is typically used by far more than one client binary, the precision of tracking the library versions and the discipline in respecting the indicated versioning information requires an extra level of strictness. Failure to notice and react to the discrepancies between the functionality provided by the different versions of a dynamic library may mean not only the malfunctioning of a single application, but sometimes chaos in the broader functionality of the operating system (file system, networking, windowing system, etc).

Gradation of Versions and their Impact on Backwards Compatibility

Not all code changes have the same impact on the module's functionality. Some of the changes are cosmetic in nature, others represent bug fixes, and still others bring in substantial changes that break away from the paradigms that existed before. The gradation of overall importance of changes is manifested in the sophisticated versioning scheme whose details deserve dedicated discussion.

Major Version Code Changes

As a rule, changes in the dynamic library code that break previously supported functionality should result with the incremented *major version* number. The symptoms of disrupted previous functionality cover a wide range of possibilities, including the following:

- A substantial change in provided runtime functionality, such as the complete elimination of a previously supported feature, substantial change of requirements for a feature to be supported, etc.

- Inability of the client binary to link against the dynamic library due to a changed ABI, such as removed functions or whole interfaces, changed exported function signatures, reordered structure or class layout, etc.

- Completely changed paradigms in maintaining the running process or changes requiring the major infrastructure changes (such as switching to a completely different type of database, starting to rely on different forms of encryption, starting to require different types of hardware, etc.).

Minor Version Code Changes

Changes in the dynamic library code that introduce new functionality without breaking the existing functionality typically result with the incremented *minor version* number. The code changes that qualify for the increment of dynamic library *minor version* numbers typically do not impose recompiling/relinking of the client binaries, nor cause substantially changed runtime behavior. The added features typically do not represent radical turns, but rather a mild enhancement of the existing assortment of available choices.

Modifications of ABI interfaces are not automatically precluded in the case of minor version increment code changes. The ABI modifications in the case of minor version changes, however, mostly mean additions of new functions, constants, and structures or classes—in other words, changes that do not impact the definition and use of the previously existing interfaces. Most importantly, the client binaries that relied on the previous version do not require rebuilding in order to use the new minor version of the dynamic library.

Patch Version

Code changes that are mostly of internal scope, which neither cause any change in the ABI interface nor bring a substantial functionality change typically qualify for the "patch" status.

Linux Dynamic Library Versioning Schemes

The Linux-specific implementation of the versioning concept will be discussed in detail, as the sophistication with which it resolves some of the most important questions related to the dynamic libraries versioning problem definitely deserves attention. Two distinct versioning schemes are currently in use: the versioning scheme based on the library's soname and the individual symbols versioning scheme.

Linux Soname-based Versioning Scheme
Linux Library Filename Carries the Version Information

As mentioned in chapter 7, the last part of the filename of a Linux dynamic library represents the library's versioning information:

```
library filename = lib + <library name> + .so + <library version information>
```

The *library version information* typically uses the format

```
dynamic library version information = <M>.<m>.<p>
```

where the M represents one or more digits indicating the library major version, the m represents one or more digits indicating the library minor version, and the p stands for one or more digits indicating the library patch (i.e., very minor change) number.

The Usual Dynamic Library Upgrade Practices

In typical real-world scenarios, arrivals of new minor versions of dynamic libraries tend to happen fairly frequently. The expectations of the minor version upgrade causing any problems are generally very low, especially if the vendor follows solid testing procedures before publishing the new code.

Most of the time, installation of the new minor version of the dynamic library should be a fairly simple and smooth procedure, such as a simple file copying of a new file.

However, regardless of how small the chances are that a new minor version breaks existing functionality, there is still a possibility that it will happen. In order to be able to elegantly step back and restore the previous version of the dynamic library which worked flawlessly, the simple file copying needs to be replaced by a bit more subtle approach.

Preamble: The Flexibility of Softlinks

By definition, a softlink is the element of the file system that carries a string containing the path to another file. In fact, we may say that softlink *points to* an existing file. In most aspects, the operating system treats the softlink as the file to which it points. The access to a softlink and redirection to the file it represents impose negligible performance penalties.

The softlink may be easily created.

```
$ ln -s <file path> <softlink path>
```

It can also be redirected to point to another file.

```
$ ln -s -f <another file> <existing softlink>
```

Finally, the softlink may be destroyed when no longer needed.

```
$ rm -rf <softlink path>
```

Preamble: Library Soname vs. Library Filename

As mentioned in the Chapter 7 discussion about the Linux library naming conventions, the library file name should adhere to the following scheme:

library filename = **lib** + <*library name*> + **.so** + <library complete version information>

The soname of the Linux dynamic library is defined as

library **soname** = lib + <*library name*> + **.so** + <(only the)library **major version** digit(s)>

Obviously, the soname is almost identical to the library filename, the only difference being that it does not carry the complete versioning information but only carries the dynamic library *major* version. As you will see, this fact plays particularly important role in the dynamic libraries versioning schemes.

Combining Softlink and Soname in the Library Upgrade Scheme

The softlink's flexibility lends itself really well to scenarios of upgrading the dynamic libraries. The following guidelines describe the procedure:

- In the same folder where the actual dynamic library filename resides, a softlink is maintained which points to the actual library file.

- Its name exactly matches the soname of the library to which it points. That way, the softlink in fact carries the library name in which the versioning information is somewhat relaxed (i.e. carries no more than the major version information).

- As a rule, the client binaries are never (i.e., only exceptionally rarely) linked against the dynamic library filename carrying the fully detailed version information. Instead, as you will see in detail shortly, the client binary build procedure is purposefully set to result with the client binary being linked against the library soname.

- The reasoning behind this decision is fairly simple: specifying the full and exact versioning information of the dynamic library would impose too many unnecessary restrictions, as it would directly prevent linking against any newer version of the same library.

Figure 10-1 illustrates the concept.

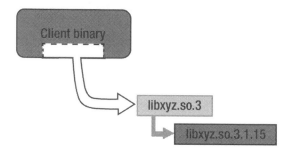

Figure 10-1. *The role of the softlink whose name matches the library's soname*

Extra Softlink Needed as Convenience for Development Scenarios

When building the client binary, you need to determine the build-time location of the dynamic library, during which you are expected to follow the rules of the "-L -l" convention. Even though it is possible to pass the exact dynamic library filename (or softlink/soname) to the linker by adding the colon character between the "-l" and the filename (-l:<filename>), such as

```
$ gcc -shared <inputs> -l:libxyz.so.1 -o <clientBinary>
```

it is the informal but well settled convention to pass only the library name deprived from any versioning information. For example,

```
$ gcc -shared <inputs> -lm -ldl -lpthread -lxml2 -lxyz -o <clientBinary>
```

indicates that the client binary requires linking with libraries whose names are libm, libdl, libpthread, libxml2, and libxyz, respectively.

For that reason, in addition to the softlink carrying the library's soname, it is typical to provide the softlink carrying just the library name plus .so file extension, as illustrated in Figure 10-2.

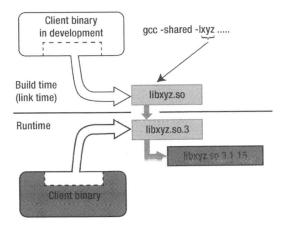

Figure 10-2. *The use of softlinks during build time vs. during runtime*

There are several ways to provide the extra softlink. The most structured way to do it is through the package deployment configuration (pkg-config). A somewhat less structured way is to do it in the deployment target of the makefile that governs the building of the dynamic library. Finally, it is always possible to create the softlink manually from the command line or by setting up a simple script to do it.

Analysis of Soname-based Versioning Scheme

The described scheme obviously combines two flexibilities: the inherent flexibility of a softlink with the soname's versioning flexibility. Here is how the two flexibilities play together in the overall scheme of things.

The Softlink's Role

Since the operating system treats the softlink as the file it points to, and inherently provides efficient dereferencing mechanisms, the loader has no particular problem connecting the client binary through the softlink with the actual library file available at runtime.

When a new version of a dynamic library arrives, it takes very little effort and time to copy its file into the same folder where the older version already resides, and to modify the softlink to point to the newer version file.

```
$ ln -s -f <new version of dynamic library file> <existing soname>
```

The benefits of this scheme are obvious:

- No need to rebuild the client binary.

- No need to erase or overwrite the current version of the dynamic library file. Both files can coexist in the same folder.

- Easy, elegant, on-the-fly setting up the client binary to use the newer version of the dynamic library.

- The ability to elegantly restore the client binary's connection with the older dynamic library version in cases when the upgrade results with the unexpected functionality.

Version Safeguarding Role of Soname

As mentioned in the previous section, not all kinds of changes in the dynamic library code will have a disruptive impact on the client binary functionality. The minor version increments are reasonably expected to not cause any major problems (such as the inability to dynamically link or run, or severe unwanted and unexpected runtime changes). The upgrades require major version increments; on the other hand, they are extremely risky proposition, and should be taken with extreme caution.

It does not require a whole lot of thinking to conclude that the soname is in fact designed to act like a fairly elastic safeguard of a kind.

- By using it as a dynamic library identifier in the process of building the client binary, you basically impose limits on the major version of the dynamic library.

  The loader is designed to be smart enough to recognize the attempt to upgrade the dynamic library to a major version different from what the soname suggests and prevent it from happening.

- By purposefully omitting the details about the minor version and patch number, you implicitly allow the minor version changes to happen without much of a hassle.

As great as this all sounds, this scheme is fairly safe only in the scenarios in which you have good reason to expect that the changes brought by the new library version will not break the overall functionality, which is the case when *at most the minor version changes*. Figure 10-3 illustrates version safeguarding role of soname.

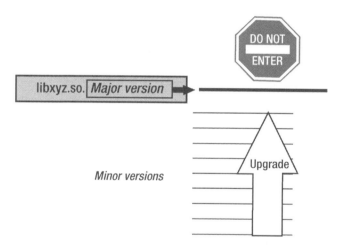

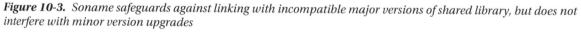

Figure 10-3. *Soname safeguards against linking with incompatible major versions of shared library, but does not interfere with minor version upgrades*

In situations when the new dynamic library features an upgraded major version, this scheme is by design prevented from running. Explaining how exactly the limitation measures work in this case requires us to dive a little bit deeper into the implementation details.

Technicalities of the Soname Implementation

As fundamentally solid as it sounds, the scheme based on using soname would not be nearly as powerful unless its implementation featured a very important facet. More specifically, *the soname gets embedded into the binaries*. The ELF format reserves the dedicated fields of the dynamic section that are used (depending on the purpose) to carry the soname information. During the linking stage, the linker takes the specified soname string and inserts it into the ELF format field of choice.

The "undercover life" of the soname starts when the linker imprints it into the dynamic library, with the purpose of declaring the library's major version. However, it does not end there. Whenever a client binary links against the dynamic library, the linker extracts the dynamic library's soname and inserts it into the client binary's file as well, albeit this time with a bit different purpose—to indicate the versioning requirements of the client binary.

Soname Embedded into the Dynamic Library File

When building a dynamic library, you can use the dedicated linker flag to specify the library soname.

```
$ gcc -shared <list of linker inputs> -Wl,-soname,<soname> -o <library filename>
```

The linker embeds the specified soname string into the DT_SONAME field of the binary, as shown in Figure 10-4.

```
milan@milan$ ls -alg
total 12
drwxrwxr-x 2 milan 4096 Dec 11 22:41 .
drwxr-xr-x 7 milan 4096 Dec 10 00:10 ..
-rw-rw-r-- 1 milan   43 Dec 11 22:40 test.c
-rw-rw-r-- 1 milan   41 Dec 11 23:01 test.h
milan@milan$ gcc -fPIC -c test.c -o test.o
milan@milan$ gcc -shared test.o -Wl,-soname,libtest.so.1 -o libtest.so.1.0.0
milan@milan$ ls -alg
total 24
drwxrwxr-x 2 milan 4096 Dec 11 22:42 .
drwxr-xr-x 7 milan 4096 Dec 10 00:10 ..
-rwxrwxr-x 1 milan 6864 Dec 11 22:42 libtest.so.1.0.0
-rw-rw-r-- 1 milan   43 Dec 11 22:40 test.c
-rw-rw-r-- 1 milan   41 Dec 11 23:01 test.h
-rw-rw-r-- 1 milan  864 Dec 11 22:41 test.o
milan@milan$ readelf -d libtest.so.1.0.0

Dynamic section at offset 0xf20 contains 21 entries:
  Tag        Type                         Name/Value
 0x00000001 (NEEDED)                     Shared library: [libc.so.6]
 0x0000000e (SONAME)                     Library soname: [libtest.so.1]
 0x0000000c (INIT)                       0x304
 0x0000000d (FINI)                       0x478
             ○
             ○
             ○
             ○
```

Figure 10-4. Soname gets embedded into the DT_SONAME field of the binary file

Soname Propagated into the Client Binary File

When client binary gets linked (either directly or through the softlink) with the dynamic library, the linker gets the dynamic library soname and inserts it into the DT_NEEDED field of the client binary, as shown in Figure 10-5.

```
milan@milan$ ln -s libtest.so.1 libtest.so
milan@milan$ ls -alg
total 32
drwxrwxr-x 3 milan 4096 Dec 11 23:23 .
drwxr-xr-x 7 milan 4096 Dec 10 00:10 ..
drwxrwxr-x 2 milan 4096 Dec 11 23:21 clientBinary
lrwxrwxrwx 1 milan   12 Dec 11 23:23 libtest.so -> libtest.so.1
lrwxrwxrwx 1 milan   16 Dec 11 23:22 libtest.so.1 -> libtest.so.1.0.0
-rwxrwxr-x 1 milan 6662 Dec 11 22:42 libtest.so.1.0.0
-rw-rw-r-- 1 milan   43 Dec 11 22:40 test.c
-rw-rw-r-- 1 milan   41 Dec 11 23:01 test.h
-rw-rw-r-- 1 milan  864 Dec 11 22:41 test.o
milan@milan$ cd clientBinary/
milan@milan:clientBinary$ ls -alg
total 12
drwxrwxr-x 2 milan 4096 Dec 11 23:20 .
drwxrwxr-x 3 milan 4096 Dec 11 23:21 ..
-rw-rw-r-- 1 milan  110 Dec 11 23:17 main.c
milan@milan:clientBinary$ gcc -I../ -c main.c -o main.o
milan@milan:clientBinary$ gcc -shared -L../ -ltest main.o -o clientBinary
milan@milan:clientBinary$ ls -alg
total 24
drwxrwxr-x 2 milan 4096 Dec 11 23:21 .
drwxrwxr-x 3 milan 4096 Dec 11 23:21 ..
-rwxrwxr-x 1 milan 6683 Dec 11 23:21 clientBinary
-rw-rw-r-- 1 milan  110 Dec 11 23:17 main.c
-rw-rw-r-- 1 milan  952 Dec 11 23:21 main.o
milan@milan:clientBinary$ readelf -d clientBinary

Dynamic section at offset 0xf18 contains 22 entries:
  Tag        Type                         Name/Value
 0x00000001 (NEEDED)                     Shared library: [libtest.so.1]
 0x00000001 (NEEDED)                     Shared library: [libc.so.6]
 0x0000000c (INIT)                       0x320
 0x0000000d (FINI)                       0x498
              ○
              ○
              ○
              ○
```

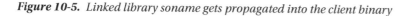

Figure 10-5. *Linked library soname gets propagated into the client binary*

That way, the versioning information carried by the soname gets propagated further, establishing firm versioning rules between all parties involved (the linker, the dynamic library file, the client binary file, and the loader).

Unlike the library filenames, which can be fairly easily modified by everybody (ranging from a younger sibling with too many fingers per brain cell and too much time all the way to malicious hackers), changing the soname value is neither a simple nor a practical task, as it requires not only modifications of the binary file but also thorough familiarity with the ELF format.

The Support from the Other Utility Programs (ldconfig)

In addition to being supported by all the necessary players in the dynamic linking scenario (i.e., the linker, the binary files, the loader), the other tools tend to support the soname concept. The ldconfig utility program is a notable example in that regard. In addition to its original scope of responsibilities, this tool has an extra "Swiss knife" feature.

When -n <directory> command line arguments are passed, the ldconfig opens up all the dynamic library files (whose names conform to the library naming convention!), extracts their soname, and for each of them creates a softlink whose name is equal to the extracted soname.

The -l <specific library file> option is even more flexible, as in this case the dynamic library filename may be absolutely any legal filename. No matter what the filename may look like (be it the fully fledged original library name with the full versioning information or a severely altered filename), the soname embedded into the specified file gets extracted and the correct softlink gets created unambiguously pointing to the library file.

To demonstrate this, a small experiment was run in which the original library name was purposefully altered. Yet the ldconfig managed to create the correct softlink, as shown in Figure 10-6.

```
milan@milan$ ls -alg
total 28
drwxrwxr-x 2 milan 4096 Dec 11 23:01 .
drwxr-xr-x 7 milan 4096 Dec 10 00:10 ..
-rwxrwxr-x 1 milan 6662 Dec 11 22:42 libtest.so.1.0.0
milan@milan$ mv libtest.so.1.0.0 purposefullyChangedName
milan@milan$ ls -alg
total 28
drwxrwxr-x 2 milan 4096 Dec 11 23:02 .
drwxr-xr-x 7 milan 4096 Dec 10 00:10 ..
-rwxrwxr-x 1 milan 6864 Dec 11 22:42 purposefullyChangedName
milan@milan$ ldconfig -l purposefullyChangedName
milan@milan$ ls -alg
total 28
drwxrwxr-x 2 milan 4096 Dec 11 23:02 .
drwxr-xr-x 7 milan 4096 Dec 10 00:10 ..
lrwxrwxrwx 1 milan   23 Dec 11 23:02 libtest.so.1 -> purposefullyChangedName
-rwxrwxr-x 1 milan 6864 Dec 11 22:42 purposefullyChangedName
milan@milan$
```

Figure 10-6. *Regardless of the library name, ldconfig extracts its soname*

Linux Symbol Versioning Scheme

In addition to controlling the versioning information of the whole dynamic library, the GNU linker supports an extra level of control over the versioning, in which the version information may be attributed to individual symbols. In this scheme, the text files known as version scripts featuring a fairly simple syntax are passed to the linker during the linking stage, which the linker inserts into the ELF sections (.gnu.version and similar ones) specialized in carrying the symbol versioning information.

The Advantage of Symbol Versioning Mechanism

The symbol versioning scheme is in many aspects more sophisticated than the soname-based versioning. A particularly interesting detail of the symbol versioning approach is that it allows a single dynamic library binary file to simultaneously carry several different versions of the same symbol. The different client binaries that need different versions of the same dynamic library will be loading the same, one and only binary file, and yet will be able to link against the symbols of a specified version.

For comparison, when the soname-based versioning method is used, in order to support several major versions of the same library, you need exactly that many different binaries (each carrying a different soname value) to be physically present on the target machine. Figure 10-7 illustrates the difference between the versioning schemes.

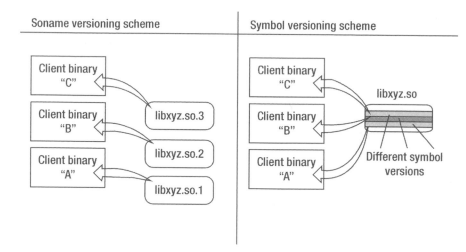

Figure 10-7. *Comparison of soname-based and symbol-based versioning schemes*

As an added bonus, due to the richness of the features supported by the script file syntax, it is also possible to control the symbol visibility (i.e., which symbols are exported by the library and which remain hidden), in the manner whose elegance and simplicity surpasses all the symbol visibility methods described so far.

Symbol Versioning Mechanisms Analysis Model

In order to fully understand the symbol versioning mechanism, it is important to define the usual use case scenarios in which it gets used.

Phase 1: Initial Version

In the beginning, let's say that a first ever published version of the dynamic library gets happily linked with the client binary "A" and everything runs great. Figure 10-8 describes this early phase of development cycle.

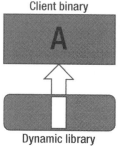

Figure 10-8. *Chronologically earliest client binary "A" links in library version 1.0.0*

This is, however, just the beginning of the story.

Phase 2: Minor Version Changes

With every day that passes, the dynamic library development progress inevitably brings changes. Even more importantly, not only does the dynamic library get changed, but also a new slew of client binaries ("B," "C," etc.) emerge, which did not exist at the time when the linking of dynamic library with the first client binary "A" happened. This stage is illustrated by Figure 10-9.

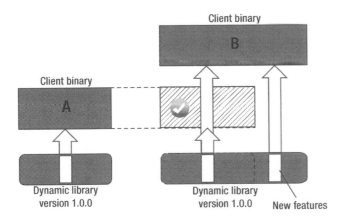

Figure 10-9. *Somewhat newer client binary "B" links in newer library version (1.1.0)*

Some of the dynamic library changes may have no implications on the already existing client binaries' functionality. Such changes are rightfully considered the *minor version* changes.

Phase 3: Major Version Changes

Occasionally, the dynamic library code changes happen to bring differences that are too radical and mean a complete breakup with what the previous library versions provided. The new client binaries ("C") created at the time of these new changes typically have no problem getting along with the new paradigm.

The older client binaries ("A" and "B"), however, may end up in the situation illustrated by Figure 10-10, which is similar to an elderly couple at a rock'n'roll wedding reception waiting forever for the band to play their favorite Glenn Miller tune.

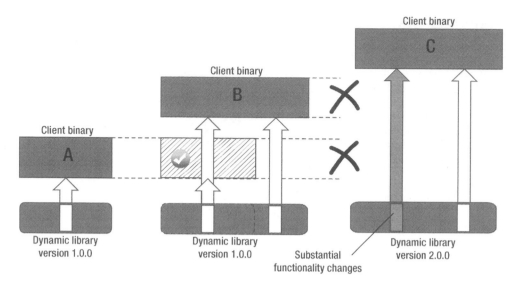

Figure 10-10. *The latest and greatest client binary "C" links in the newest dynamic library version (2.0.0), which is incompatible for use by the older client binaries "A" and "B"*

The task of software developers is to make the transition of the functionality upgrades as smooth as possible. Breaking the compatibility with the existing infrastructure is seldom a wise move. The more the library is popular among the developers, the less recommended it is to break away from the library's expected functionality. The true solution to the problem is that the new dynamic library keeps providing for both older and newer functionality versions, at least for some time. This idea is illustrated in Figure 10-11.

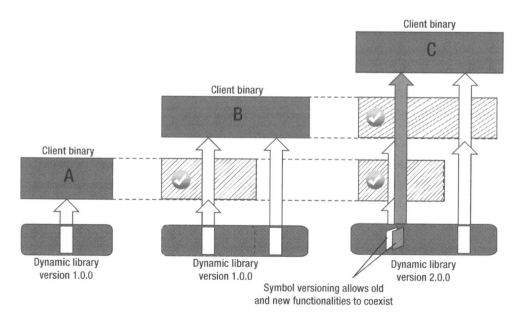

Figure 10-11. *Symbol versioning resolves incompatibility issues*

The Basic Implementation Ingredients

The symbol versioning scheme is implemented by combining the linker version script with the .symver assembler directive, both of which will be elaborated in detail next.

Linker Version Script

The most basic implementation of the symbol visibility control mechanism is based on the GNU linker reading in the symbol version information supplied in the form of the **version script** text file.

Let's start a simple demo from the example of a simple dynamic library (libsimple.so) which features the three functions shown in Listing 10-1.

Listing 10-1. simple.c

```
int first_function(int x)
{
    return (x+1);
}

int second_function(int x)
{
    return (x+2);
}

int third_function(int x)
{
    return (x+3);
}
```

Let's say now that you want the first two library functions (but not the third one!) to carry the versioning information. The way to specify the symbol version is to create a fairly simple version script file, which may look somewhat like the code in Listing 10-2.

Listing 10-2. simpleVersionScript

```
LIBSIMPLE_1.0 {
    global:
        first_function; second_function;

    local:
        *;
};
```

Finally, let's now build the dynamic library. The version script filename may be conveniently passed to the linker by using the dedicated linker flag, like so:

```
$ gcc -fPIC -c simple.c
$ gcc -shared simple.o -Wl,--version-script,simpleVersionScript -o libsimple.so.1.0.0
```

The linker extracts the information from the script file and embeds it into the dedicated ELF format section dedicated to versioning. More information on how the symbol versioning information gets embedded into the ELF binary files will follow shortly.

199

.symver Assembler Directive

Unlike the version script file that represents the "bread and butter" of the symbol versioning concept, used in all phases and all scenarios, the symbol versioning paradigm relies on another ingredient—the .symver assembler directive—to resolve the tough corner cases.

Let's assume a scenario of major version changes in which a function signature did not change between the versions, but the underlying functionality has changed quite a bit. Further, there's a function that originally used to return a number of linked elements, but in the latest version was redesigned to return the total number of bytes occupied by the linked list (or vice versa). See Listing 10-3.

Listing 10-3. Example of substantially different implementations of the same function which qualify for different major versions

```
// VERSION 1.0:
unsigned long list_occupancy(struct List* pStart)
{
    // here we scan the list, and return the number of elements
    return nElements;
}


// VERSION 2.0:
unsigned long list_occupancy(struct List* pStart)
{
    // here we scan the list, but now return the total number of bytes
    return nElements*sizeof(struct List);
}
```

Obviously, the clients of the library's first version will face problems, as the value returned by function will no longer match what is expected.

As stated previously, the credo of this versioning technique is to provide the different versions of the same symbol in the same binary file. Nicely said, but how to do it? An attempt to build both function versions will result in the linker reporting the duplicate symbols. Fortunately, the GCC compiler supports the custom .symver assembler directive, which helps alleviate the problem (see Listing 10-4).

Listing 10-4. The same pair of different versions of the function featured in Listing 10-3, this time with properly applied symbol versioning

```
__asm__(".symver list_occupancy_1_0, list_occupancy@MYLIBVERSION_1.0");
unsigned long list_occupancy_1_0(struct List* pStart)
{
    // here we scan the list, and return the number of elements
    return nElements;
}
                                    // default symbol version indicated by the additional "@"
                                    //                    |
                                    //                    v
__asm__(".symver list_occupancy_2_0, list_occupancy@@MYLIBVERSION_2.0");
unsigned long list_occupancy_2_0(struct List* pStart)
{
    // here we scan the list, but now return the total number of bytes
    return nElements*sizeof(struct List);
}
```

How Does This Scheme Work?

In order to eliminate the linker facing the duplicate symbols problem, you can create different names for different versions of the same function, which will be used for internal purposes only (i.e. will not be exported). These two functions are list_occupancy_1_0 and list_occupancy_2_0.

From the outer world perspective, however, the linker will create the symbol featuring the expected function name (i.e. list_occupancy()), albeit decorated with the appropriate symbol version information, appearing in the two different versions: list_occupancy@MYLIBVERSION_1.0 *and* list_occupancy@MYLIBVERSION_2.0.

As a consequence, both the older and the newer client binary will be able to identify the symbol they expect. The older client binary will be happy to see that the symbol list_occupancy@MYLIBVERSION_1.0 exists. Its calls to this intermediary function symbol will be internally routed to the right place—to the list_occupancy_1_0() dynamic library function, which is the *real* symbol.

Finally, the brand new client binaries, which do not particularly care about the previous versions history, will choose the default symbol, indicated by the extra @ character in the name (in this case, list_occupancy@@MYLIBVERSION_2.0).

Sample Project Analysis: Phase 1 (Initial Version)

Now that you understand how the basic implementation ingredients (version script and/or .symver assembler directive) work, it's time to take a closer look at a real example. To illustrate the important points, let's go back to the original example used to illustrate the linker version script (i.e. the library libsimple.so featuring three functions, the first two of which will be subject to symbol versioning). In order to make the demo more convincing, some printf's will be added to the original code; see Listing 10-5 through Listing 10-8.

Listing 10-5. simple.h

```
#pragma once
int first_function(int x);
int second_function(int x);
int third_function(int x);
```

Listing 10-6. simple.c

```
#include <stdio.h>
#include "simple.h"

int first_function(int x)
{
    printf(" lib: %s\n", __FUNCTION__);
    return (x+1);
}

int second_function(int x)
{
    printf(" lib: %s\n", __FUNCTION__);
    return (x+2);
}

int third_function(int x)
{
    printf(" lib: %s\n", __FUNCTION__);
    return (x+3);
}
```

Listing 10-7. simpleVersionScript

```
LIBSIMPLE_1.0 {
    global:
        first_function; second_function;

    local:
        *;
};
```

Listing 10-8. build.sh

```
gcc -Wall -g -O0 -fPIC -c simple.c
gcc -shared simple.o -Wl,--version-script,simpleVersionScript -o libsimple.so.1.0.0
```

Now that the library is built, let's take a closer look at how the ELF format supports the symbol versioning concept.

ELF Format Support

The section analysis of the library file indicates that there are three sections of fairly similar names that are used to carry the version information, as shown in Figure 10-12.

```
milan@milan$ readelf -S libsimple.so
There are 35 section headers, starting at offset 0x154c:
```

Section Headers:

[Nr]	Name	Type	Addr	Off	Size	ES	Flg	Lk	Inf	Al
[0]		NULL	00000000	000000	000000	00		0	0	0
[1]	.note.gnu.build-i	NOTE	00000114	000114	000024	00	A	0	0	4
[2]	.gnu.hash	GNU_HASH	00000138	000138	00002c	04	A	3	0	4
[3]	.dynsym	DYNSYM	00000164	000164	000080	10	A	4	1	4
[4]	.dynstr	STRTAB	000001e4	0001e4	000098	00	A	0	0	1
[5]	.gnu.version	VERSYM	0000027c	00027c	000010	02	A	3	0	2
[6]	.gnu.version_d	VERDEF	0000028c	00028c	000038	00	A	4	2	4
[7]	.gnu.version_r	VERNEED	000002c4	0002c4	000030	00	A	4	1	4

Figure 10-12. *ELF format support for versioning information*

Invoking the readelf utility with the -V command line argument provides a report about the contents of these sections in a particularly neat way, as shown in Figure 10-13.

```
milan@milan$ readelf -V libsimple.so

Version symbols section '.gnu.version' contains 8 entries:
 Addr: 000000000000027c  Offset: 0x00027c  Link: 3 (.dynsym)
  000:   0 (*local*)        3 (GLIBC_2.1.3)   4 (GLIBC_2.0)     0 (*local*)
  004:   0 (*local*)        2 (LIBSIMPLE_1.0)  2 (LIBSIMPLE_1.0)  2 (LIBSIMPLE_1.0)

Version definition section '.gnu.version_d' contains 2 entries:
  Addr: 0x000000000000028c  Offset: 0x00028c  Link: 4 (.dynstr)
  000000: Rev: 1  Flags: BASE   Index: 1  Cnt: 1  Name: libsimple.so.1.0.0
  0x001c: Rev: 1  Flags: none  Index: 2  Cnt: 1  Name: LIBSIMPLE_1.0

Version needs section ' gnu version_r' contains 1 entries:
 Addr: 0x00000000000002c4  Offset: 0x0002c4  Link: 4 (.dynstr)
  000000: Version: 1  File: libc.so.6  Cnt: 2
  0x0010:   Name: GLIBC_2.0  Flags: none  Version: 4
  0x0020:   Name: GLIBC_2.1.3  Flags: none  Version: 3
milan@milan$
```

Figure 10-13. *Using readelf to list contents of version-related sections*

It becomes apparent that the

- .gnu.version_d section describes the versioning information **defined** in this particular library (hence the appendix "_d" in the section name).

- .gnu.version_r section describes the versioning information of the other libraries, which is **referenced** by this library (hence the appendix "_r" in the section name).

- .gnu_version section provides the summary list of all version information related to the library.

It is interesting at this point to verify whether the version information got associated with the symbols specified in the version script.

Of all the available ways (nm, objdump, readelf) to examine the binary file symbols, it is again the readelf utility that provides the answer in the nicest form in which the association of symbols with the specified version information becomes apparent, as illustrated in Figure 10-14.

```
milan@milan$ readelf --symbols libsimple.so | grep function
     6: 00000488    44 FUNC    GLOBAL DEFAULT    12 second_function@@LIBSIMPLE_1.0
     7: 0000045c    44 FUNC    GLOBAL DEFAULT    12 first_function@@LIBSIMPLE_1.0
    52: 000004b4    44 FUNC    LOCAL  DEFAULT    12 third_function
    64: 0000045c    44 FUNC    GLOBAL DEFAULT    12 first_function
    66: 00000488    44 FUNC    GLOBAL DEFAULT    12 second_function
milan@milan$ readelf --dyn-syms libsimple.so | grep function
     6: 00000488    44 FUNC    GLOBAL DEFAULT    12 second_function@@LIBSIMPLE_1.0
     7: 0000045c    44 FUNC    GLOBAL DEFAULT    12 first_function@@LIBSIMPLE_1.0
milan@milan$
```

Figure 10-14. *Using readelf to print symbol versioning information*

Clearly, the versioning information specified in the version script and passed to the linker found its way into the binary file and certainly became the attribute of the symbols intended for versioning.

As an interesting side note, the disassembling of the binary file, however, shows that there is no such thing as first_function@@LIBVERSIONDEMO_1.0. All you can find is the symbol of the real first_function. The disassembling at runtime (by running gdb) shows the same thing.

Obviously, the exported symbol decorated with the symbol versioning information is a kind of fiction (useful, but still a fiction), whereas the only thing that counts in the end is the symbol of the real, existing function.

Propagation of Version Symbol Information to the Client Binaries

Another round of interesting findings happens when you examine the client binaries that link against your symbol-versioned dynamic library. In order to explore the symbol versioning in that particular direction, let's create a simple demo application that references the versioned symbols; see Listing 10-9.

Listing 10-9. main.c

```c
#include <stdio.h>
#include "simple.h"

int main(int argc, char* argv[])
{
    int nFirst  = first_function(1);
    int nSecond = second_function(2);
    int nRetValue = nFirst + nSecond;
    printf("first(1) + second(2) = %d\n", nRetValue);
    return nRetValue;
}
```

Let's now build it.

```
$ gcc -g -O0 -c -I../sharedLib main.c
$ gcc main.o -Wl,-L../sharedLib -lsimple \
          -Wl,-R../sharedLib -o firstDemoApp
```

Please notice that in order to exercise solely the symbol versioning mechanism, specifying the library soname has been purposefully omitted.

It does not come as a big surprise that the demo app, being an ELF binary file, also carries the version related sections (as shown by the section inspection illustrated in the Figure 10-15).

```
milan@milan$ readelf -S ./firstDemoApp
There are 36 section headers, starting at offset 0x1454:

Section Headers:
  [Nr] Name              Type           Addr     Off    Size   ES Flg Lk Inf Al
  [ 0]                   NULL           00000000 000000 000000 00      0   0  0
  [ 1] .interp           PROGBITS       08048154 000154 000013 00   A  0   0  1
  [ 2] .note.ABI-tag     NOTE           08048168 000168 000020 00   A  0   0  4
  [ 3] .note.gnu.build-i NOTE           08048188 000188 000024 00   A  0   0  4
  [ 4] .gnu.hash         GNU_HASH       080481ac 0001ac 000020 04   A  5   0  4
  [ 5] .dynsym           DYNSYM         080481cc 0001cc 000080 10   A  6   1  4
  [ 6] .dynstr           STRTAB         0804824c 00024c 0000a7 00   A  0   0  1
  [ 7] .gnu.version      VERSYM         080482f4 0002f4 000010 02   A  5   0  2
  [ 8] .gnu.version_r    VERNEED        08048304 000304 000040 00   A  6   2  4
```

Figure 10-15. *Demo application also features versioning-related sections*

It is far more important that the demo dynamic library's symbol version information was ingested by the client binary through the process of linking, as shown in Figure 10-16.

```
milan@milan$ readelf -V ./firstDemoApp

Version symbols section '.gnu.version' contains 8 entries:
 Addr: 00000000080482f4  Offset: 0x0002f4  Link: 5 (.dynsym)
   000:   0 (*local*)        2 (GLIBC_2.0)      3 (LIBSIMPLE_1.0)   3 (LIBSIMPLE_1.0)
   004:   0 (*local*)        2 (GLIBC_2.0)      0 (*local*)         1 (*global*)

Version needs section '.gnu.version_r' contains 2 entries:
 Addr: 0x0000000008048304  Offset: 0x000304  Link: 6 (.dynstr)
   000000: Version: 1  File: libsimple.so  Cnt: 1
   0x0010:   Name: LIBSIMPLE_1.0  Flags: none  Version: 3
   0x0020: Version: 1  File: libc.so.6  Cnt: 1
   0x0030:   Name: GLIBC_2.0  Flags: none  Version: 2
milan@milan$
```

Figure 10-16. *Client binary "ingests" the symbol versioning info from library it linked in*

Exactly like it happens in the previously described soname-based versioning scenario, the symbol versioning mechanism also gets passed from the dynamic library to its client binary. This way, a form of contract between the client binary and the dynamic library versioning has been established.

Why is this important? From the moment in which the linking of the client binary with the dynamic library happened, the dynamic library code may pass through a multitude of changes and accordingly through the multitude of minor and major versions.

Regardless of the dynamic library changes, its client binary will keep carrying on the versioning information that was present at the time of the linking. If exactly that version (and of course exactly the functionality associated with that particular version) happens to be missing, the broken backwards compatibility will be strongly indicated.

Before advancing further, let's make sure that your versioning scheme does not prevent the app from running. The simple experiment is shown in Figure 10-17.

```
milan@milan$ ./firstDemoApp
 lib: first_function
 lib: second_function
first(1) + second(2) = 6
milan@milan$
```

Figure 10-17. *Versioning scheme working correctly*

Sample Project Analysis: Phase 2 (Minor Version Changes)

Once you understand the basics of how the symbol versioning scheme operates, it's time to simulate the scenario in which the dynamic library development results with the non-disruptive changes (i.e., minor version) changes. In the attempt to simulate the real-life scenarios, the following steps will be taken:

- You will modify the dynamic library by adding a few more functions. Only one of the newly added functions will be exported. The versioning script will be enriched by the extra item announcing the LIBSIMPLE_1.1 minor version upgrade.

- The original client binary (i.e., the initial simple demo application) will be purposefully left untouched. By not rebuilding it, it will perfectly mimic the legacy application, built at the time when the dynamic library featured the initial version 1.0.

- The new client binary (another simple demo application) will be created and linked against the updated dynamic library. This way, it will serve as an example of a brand new client binary, created at the time of the latest and greatest dynamic library version 1.1, unaware of any of the previous library versions.

- To simplify the demo, its code will not be significantly different from the original simple demo application. The most notable difference is that it will call the new ABI function, which did not exist prior to the latest 1.1 version.

Listings 10-10 and 10-11 show what the source file of the modified dynamic library looks like now.

Listing 10-10. simple.h

```
#pragma once

int first_function(int x);
int second_function(int x);
int third_function(int x);

int fourth_function(int x);
int fifth_function(int x);
```

Listing 10-11. simple.c

```
#include <stdio.h>
#include "simple.h"

int first_function(int x)
{
    printf(" lib: %s\n", __FUNCTION__);
    return (x+1);
}

int second_function(int x)
{
    printf(" lib: %s\n", __FUNCTION__);
    return (x+2);
}

int third_function(int x)
{
    printf(" lib: %s\n", __FUNCTION__);
    return (x+3);
}

int fourth_function(int x) // exported in version 1.1
{
    printf(" lib: %s\n", __FUNCTION__);
    return (x+4);
}
```

```
int fifth_function(int x)
{
    printf(" lib: %s\n", __FUNCTION__);
    return (x+5);
}
```

Listing 10-12 shows how the version script will look after the changes.

Listing 10-12. simpleVersionScript

```
LIBSIMPLE_1.0 {
    global:
        first_function; second_function;

    local:
        *;
};

LIBSIMPLE_1.1 {
    global:
        fourth_function;

    local:
        *;
};
```

The new demo application source file will look like Listing 10-13.

Listing 10-13. main.c

```
#include <stdio.h>
#include "simple.h"

int main(int argc, char* argv[])
{
    int nFirst  = first_function(1);
    int nSecond = second_function(2);
    int nFourth = fourth_function(4);
    int nRetValue = nFirst + nSecond + nFourth;
    printf("first(1) + second(2) + fourth(4) = %d\n", nRetValue);
    return nRetValue;
}
```

Let's now build it.

```
$ gcc -g -O0 -c -I../sharedLib main.c
$ gcc main.o -Wl,-L../sharedLib -lsimple \
            -Wl,-R../sharedLib -o newerApp
```

Let's now take a closer look at the effects of this little versioning adventure, which perfectly mimics the real life scenario happening when dynamic library minor versions gets upgraded.

First, as shown in Figure 10-18, the versioning information now features not only the original version (1.0) but also the newest version (1.1)

```
milan@milan$ readelf -V libsimple.so

Version symbols section '.gnu.version' contains 10 entries:
 Addr: 00000000000002c2  Offset: 0x0002c2  Link: 3 (.dynsym)
  000:   0 (*local*)       4 (GLIBC_2.1.3)   5 (GLIBC_2.0)      0 (*local*)
  004:   0 (*local*)       3 (LIBSIMPLE 1.1)  2 (LIBSIMPLE_1.0)  2 (LIBSIMPLE_1.0)
  008:   2 (LIBSIMPLE_1.0)  3 (LIBSIMPLE 1.1)

Version definition section '.gnu.version_d' contains 3 entries:
  Addr: 0x00000000000002d8  Offset: 0x0002d8  Link: 4 (.dynstr)
  000000: Rev: 1  Flags: BASE  Index: 1  Cnt: 1  Name: libsimple.so.1.0.0
  0x001c: Rev: 1  Flags: none  Index: 2  Cnt: 1  Name: LIBSIMPLE_1.0
  0x0038: Rev: 1  Flags: none  Index: 3  Cnt: 1  Name: LIBSIMPLE 1.1

Version needs section '.gnu.version_r' contains 1 entries:
 Addr: 0x000000000000032c  Offset: 0x00032c  Link: 4 (.dynstr)
  000000: Version: 1  File: libc.so.6  Cnt: 2
  0x0010:   Name: GLIBC_2.0 Flags: none  Version: 5
  0x0020:   Name: GLIBC_2.1.3 Flags: none  Version: 4
milan@milan$
```

Figure 10-18. *Complete versioning information ingested by the client binary*

The set of exported symbols is now comprised of both version 1.0 and version 1.1 symbols, as shown in Figure 10-19.

```
milan@milan$ readelf --dyn-sym libsimple.so

Symbol table '.dynsym' contains 10 entries:
   Num:    Value  Size Type    Bind   Vis      Ndx Name
     0: 00000000     0 NOTYPE  LOCAL  DEFAULT  UND
     1: 00000000     0 FUNC    WEAK   DEFAULT  UND __cxa_finalize@GLIBC_2.1.3 (4)
     2: 00000000     0 FUNC    GLOBAL DEFAULT  UND puts@GLIBC_2.0 (5)
     3: 00000000     0 NOTYPE  WEAK   DEFAULT  UND __gmon_start__
     4: 00000000     0 NOTYPE  WEAK   DEFAULT  UND _Jv_RegisterClasses
     5: 00000550    44 FUNC    GLOBAL DEFAULT   12 fourth_function@@LIBSIMPLE 1.1
     6: 00000000     0 OBJECT  GLOBAL DEFAULT  ABS LIBSIMPLE_1.0
     7: 000004f8    44 FUNC    GLOBAL DEFAULT   12 second_function@@LIBSIMPLE_1.0
     8: 000004cc    44 FUNC    GLOBAL DEFAULT   12 first_function@@LIBSIMPLE_1.0
     9: 00000000     0 OBJECT  GLOBAL DEFAULT  ABS LIBSIMPLE_1.1
milan@milan$
```

Figure 10-19. *Symbols of different versions present in the shared library*

Let's see now how the things look with the newer, more modern client binary (newerApp) built for the first time after version 1.1 came out. As illustrated in Figure 10-20, the linker read out the information about all the versions supported by the dynamic library and inserted it into the newer app's client binary.

```
milan@milan$ readelf -V ./newerApp

Version symbols section '.gnu.version' contains 9 entries:
 Addr: 0000000008048322  Offset: 0x000322  Link: 5 (.dynsym)
   000:    0 (*local*)        2 (GLIBC_2.0)      3 (LIBSIMPLE_1.0)    4 (LIBSIMPLE 1.1)
   004:    3 (LIBSIMPLE_1.0)  0 (*local*)        2 (GLIBC_2.0)        0 (*local*)
   008:    1 (*global*)

Version needs section '.gnu.version_r' contains 2 entries:
 Addr: 0x0000000008048334  Offset: 0x000334  Link: 6 (.dynstr)
  000000: Version: 1  File: libsimple.so  Cnt: 2
  0x0010:   Name: LIBSIMPLE 1.1  Flags: none  Version: 4
  0x0020:   Name: LIBSIMPLE_1.0  Flags: none  Version: 3
  0x0030: Version: 1  File: libc.so.6  Cnt: 1
  0x0040:   Name: GLIBC_2.0  Flags: none  Version: 2
milan@milan$
```

Figure 10-20. *Newer client binary ingested complete versioning info (both old and newer symbol versions)*

The list of dynamic library symbols on whose presence the client binary counts on at runtime contains the symbols of both versions. Figure 10-21 illustrates the point.

```
milan@milan$ readelf --dyn-syms ./newerApp

Symbol table '.dynsym' contains 9 entries:
   Num:    Value  Size Type    Bind   Vis      Ndx Name
     0: 00000000     0 NOTYPE  LOCAL  DEFAULT  UND
     1: 00000000     0 FUNC    GLOBAL DEFAULT  UND printf@GLIBC_2.0 (2)
     2: 00000000     0 FUNC    GLOBAL DEFAULT  UND second_function@LIBSIMPLE 1.0 (3)
     3: 00000000     0 FUNC    GLOBAL DEFAULT  UND fourth_function@LIBSIMPLE_1.1 (4)
     4: 00000000     0 FUNC    GLOBAL DEFAULT  UND first_function@LIBSIMPLE_1.0 (3)
     5: 00000000     0 NOTYPE  WEAK   DEFAULT  UND __gmon_start__
     6: 00000000     0 FUNC    GLOBAL DEFAULT  UND __libc_start_main@GLIBC_2.0 (2)
     7: 00000000     0 NOTYPE  WEAK   DEFAULT  UND _Jv_RegisterClasses
     8: 0804864c     4 OBJECT  GLOBAL DEFAULT   15 _IO_stdin_used
milan@milan$
```

Figure 10-21. *Symbols of all versions ingested from the shared library*

Now, in order to verify that the addition of the new functionality and modified versioning information works as expected, you can try to run both the old and new application. As shown in Figure 10-22, running the old app will prove that the new minor version of the dynamic library did not bring any unpleasant surprises.

```
milan@milan$ ./newerApp
 lib: first_function
 lib: second_function
 lib: fourth_function
first(1) + second(2) + fourth(4) = 14
milan@milan$ ./firstDemoApp
 lib: first_function
 lib: second_function
first(1) + second(2) = 6
milan@milan$
```

Figure 10-22. *Both older and newer app link the same library, but use the symbols of different versions*

Sample Project Analysis: Phase 3 (Major Version Changes)

In the previously analyzed examples I've covered cases in which the new code changes generally did not impact how the clients used the existing code base. Such code changes are rightfully recognized as minor version increments.

I will not try to cover much more dramatic situations, in which the code changes seriously undermine the way of how the clients used the code, thus clearly falling into the major version increments category.

The Case of Changed ABI Function Behavior

Potentially the most unpleasant code changes happen when seemingly nothing happens with the dynamic library symbols (i.e., the functions do not have their prototypes altered, and/or the structures do not have their layout changed), but the underlying meaning of what the functions do with the data—and most importantly, the values they return—does change.

Imagine for a moment that you have a function that used to return the time value in milliseconds. One fine day, the developers figured out that the millisecond as a measure was not precise enough, and decide to return the value in nanoseconds instead (which is 1,000 times larger).

This scenario is what we will use as the topic of the next example; I'll show how problems of this nature may be solved by clever use of the symbol versioning mechanism. (I do agree that example is a bit childish/fantastic/naive. Indeed, there are about a million ways how the chaos ensuing from this change may be avoided. For example, you could introduce a new ABI function with the word "nanoseconds" in the name, which would return time in nanoseconds. Even then, an example like this one is sufficiently good for the demo purposes.)

Back to the topic, let's assume that the demo dynamic library export header hasn't changed at all, so the function prototypes are unchanged. However, the most recent design requirements dictate that the first_function() from now on needs to return a value different from what it used to return.

```
int first_function(int x)
{
    printf(" lib: %s\n", __FUNCTION__);
    return 1000*(x+1);
}
```

Needless to say, this kind of change is bound to wreak havoc with the existing client binaries. Their existing code infrastructure simply does not expect a value of that order of magnitude. It is possible that stepping out of the bounds of arrays causes an exception. In graph-plotting scenarios, the value would be way out of bounds, etc.

So now you need a way to make sure that the old customers get the usual treatment (i.e., the existing client binaries' calls to first_function() return the value it used to), whereas the new customers get the benefit of a new design.

The only problem is that you have to resolve the conflict; the same function name must be used in two substantially different scenarios. Fortunately, the symbol versioning mechanism proves it is able to handle problems of this kind.

As a first step, you will modify the version script to indicate support for the new major version; see Listing 10-14.

Listing 10-14. simpleVersionScript

```
LIBSIMPLE_1.0 {
    global:
        first_function; second_function;
    local:
        *;
};
```

```
LIBSIMPLE_1.1 {
    global:
        fourth_function;
    local:
        *;
};

LIBSIMPLE_2.0 {
    global:
        first_function;
    local:
        *;
};
```

Next, you will apply the recipe based on using the .symver assembler directive, as shown in Listing 10-15.

Listing 10-15. simple.c (only the changes shown here)

```
...
__asm__(".symver first_function_1_0,first_function@LIBSIMPLE_1.0");
int first_function_1_0(int x)
{
    printf(" lib: %s\n", __FUNCTION__);
    return (x+1);
}

__asm__(".symver first_function_2_0,first_function@@LIBSIMPLE_2.0");
int first_function_2_0(int x)
{
    printf(" lib: %s\n", __FUNCTION__);
    return 1000*(x+1);
}
...
```

As shown in Figure 10-23, the dynamic library now features one extra piece of versioning information.

```
milan@milan$ readelf -V libsimple.so

Version symbols section '.gnu.version' contains 12 entries:
 Addr: 00000000000002fe  Offset: 0x0002fe  Link: 3 (.dynsym)
  000:   0 (*local*)        5 (GLIBC_2.0)      6 (GLIBC_2.1.3)    0 (*local*)
  004:   0 (*local*)        3 (LIBSIMPLE_1.1)  2 (LIBSIMPLE_1.0)  4 (LIBSIMPLE 2.0)
  008:   2 (LIBSIMPLE_1.0)  4 (LIBSIMPLE 2.0)  3 (LIBSIMPLE_1.1)  2h(LIBSIMPLE_1.0)

Version definition section '.gnu.version_d' contains 4 entries:
  Addr: 0x0000000000000318  Offset: 0x000318  Link: 4 (.dynstr)
  000000: Rev: 1  Flags: BASE   Index: 1  Cnt: 1  Name: libsimple.so.1.0.0
  0x001c: Rev: 1  Flags: none  Index: 2  Cnt: 1  Name: LIBSIMPLE_1.0
  0x0038: Rev: 1  Flags: none  Index: 3  Cnt: 1  Name: LIBSIMPLE_1.1
  0x0054: Rev: 1  Flags: none  Index: 4  Cnt: 1  Name: LIBSIMPLE_2.0

Version needs section '.gnu.version_r' contains 1 entries:
 Addr: 0x0000000000000388  Offset: 0x000388  Link: 4 (.dynstr)
  000000: Version: 1  File: libc.so.6  Cnt: 2
  0x0010:   Name: GLIBC_2.1.3  Flags: none  Version: 6
  0x0020:   Name: GLIBC_2.0  Flags: none  Version: 5
milan@milan$
```

Figure 10-23. *The latest and greatest library version contains all the symbol versions*

Interestingly, as shown in Figure 10-24, it looks like the .symver directive actually did its magic.

```
milan@milan$ readelf --dyn-syms libsimple.so

Symbol table '.dynsym' contains 12 entries:
   Num:    Value  Size Type    Bind   Vis      Ndx Name
     0: 00000000     0 NOTYPE  LOCAL  DEFAULT  UND
     1: 00000000     0 FUNC    GLOBAL DEFAULT  UND printf@GLIBC_2.0 (5)
     2: 00000000     0 FUNC    WEAK   DEFAULT  UND __cxa_finalize@GLIBC_2.1.3 (6)
     3: 00000000     0 NOTYPE  WEAK   DEFAULT  UND __gmon_start__
     4: 00000000     0 NOTYPE  WEAK   DEFAULT  UND _Jv_RegisterClasses
     5: 000005fa    54 FUNC    GLOBAL DEFAULT   12 fourth_function@@LIBSIMPLE_1.1
     6: 00000000     0 OBJECT  GLOBAL DEFAULT  ABS LIBSIMPLE_1.0
     7: 00000000     0 OBJECT  GLOBAL DEFAULT  ABS LIBSIMPLE_2.0
     8: 0000058e    54 FUNC    GLOBAL DEFAULT   12 second_function@@LIBSIMPLE_1.0
     9: 00000552    60 FUNC    GLOBAL DEFAULT   12 first_function@@LIBSIMPLE_2.0
    10: 00000000     0 OBJECT  GLOBAL DEFAULT  ABS LIBSIMPLE_1.1
    11: 0000051c    54 FUNC    GLOBAL DEFAULT   12 first_function@LIBSIMPLE_1.0
milan@milan$ nm libsimple.so | grep function
00000630 t fifth_function
00000552 T first_function@@LIBSIMPLE_2.0
0000051c T first_function@LIBSIMPLE_1.0
0000051c t first_function_1_0
00000552 t first_function_2_0
000005fa T fourth_function
0000058e T second_function
000005c4 t third_function
milan@milan$
```

Figure 10-24. *Both versions of first_function() exist*

The ultimate effect of the whole .symver scheme is the magic of exporting two versions of the first_function() symbol, despite the fact that a function of such name no longer exists because it was replaced by first_function_1_0() and first_function_2_0().

In order to clearly show the implementation differences, you will create the new application whose source does not differ from previous version (see Listing 10-16).

Listing 10-16. main.c

```
#include <stdio.h>
#include "simple.h"

int main(int argc, char* argv[])
{
    int nFirst  = first_function(1); // seeing 1000 times larger return value will be fun!
    int nSecond = second_function(2);
    int nFourth  = fourth_function(4);
    int nRetValue = nFirst + nSecond + nFourth;
    printf("first(1) + second(2) + fourth(4) = %d\n", nRetValue);
    return nRetValue;
}
```

The new app name will be chosen accordingly:

```
$ gcc -g -O0 -c -I../sharedLib main.c
$ gcc main.o -Wl,-L../sharedLib -lsimple \
            -Wl,-R../sharedLib -o ver2PeerApp
```

The runtime comparison will clearly show that the old clients will have their functionality unaffected by the major version changes. The contemporary app, however, will count on the new functionality brought by version 2.0. Figure 10-25 summarizes the point.

```
milan@milan$ ./firstDemoApp
 lib: first_function_1_0
 lib: second_function
first(1) + second(2) = 6
milan@milan$ ./newerApp
 lib: first_function_1_0
 lib: second_function
 lib: fourth_function
first(1) + second(2) + fourth(4) = 14
milan@milan$ ./ver2PeerApp
 lib: first_function_2_0
 lib: second_function
 lib: fourth_function
first(1) + second(2) + fourth(4) = 2012
milan@milan$
```

Figure 10-25. *Three apps (each of which rely on different symbol versions of the same dynamic library) run as intended*

The Case of Changed ABI Function Prototype

The previously described case is a bit bizarre. Due to the numerous ways it can be avoided, the chances of it happening in real life are fairly low. From an education standpoint, however, it is precious as the procedure of fixing such problem is the simplest as can be.

A far more usual case that falls under the major version code changes is when the signature of a function needs to be changed. For example, let's assume that for the new use case scenarios the first_function() needs to accept an additional input argument.

```
int first_function(int x, int normfactor);
```

Obviously, you now need to support the functions of the same name but of different signatures. In order to demonstrate this problem, let's create another version, shown in Listing 10-17.

Listing 10-17. simpleVersionScript

```
LIBSIMPLE_1.0 {
    global:
        first_function; second_function;
    local:
        *;
};

LIBSIMPLE_1.1 {
    global:
        fourth_function;
    local:
        *;
};

LIBSIMPLE_2.0 {
    global:
        first_function;
    local:
        *;
};

LIBSIMPLE_3.0 {
    global:
        first_function;
    local:
        *;
};
```

In general, the solution for this problem does not substantially differ from the previous case, as the recipe based on the .symver assembler directive will be used much in the same way as in the previous example (see Listing 10-18).

Listing 10-18. simple.c (only the changes shown here)

```
__asm__(".symver first_function_1_0,first_function@LIBSIMPLE_1.0");
int first_function_1_0(int x)
{
    printf(" lib: %s\n", __FUNCTION__);
    return (x+1);
}

__asm__(".symver first_function_2_0,first_function@LIBSIMPLE_2.0");
int first_function_2_0(int x)
{
    printf(" lib: %s\n", __FUNCTION__);
    return 1000*(x+1);
}

__asm__(".symver first_function_3_0,first_function@@LIBSIMPLE_3.0");
int first_function_3_0(int x, int normfactor)
{
    printf(" lib: %s\n", __FUNCTION__);
    return normfactor*(x+1);
}
```

The most substantial difference, however, is that the export header must be modified, as shown in Listing 10-19.

Listing 10-19. simple.h

```
#pragma once

// defined when building the latest client binary
#ifdef SIMPLELIB_VERSION_3_0
int first_function(int x, int normfactor);
#else
int first_function(int x);
#endif // SIMPLELIB_VERSION_3_0

int second_function(int x);
int third_function(int x);

int fourth_function(int x);
int fifth_function(int x);
```

Only the client binary built with the SIMPLELIB_VERSION_3_0 preprocessor constant passed to the compiler will include the new first_function() prototype.

```
$ gcc -g -O0 -c -DSIMPLELIB_VERSION_3_0 -I../sharedLib main.c
$ gcc main.o -Wl,-L../sharedLib -lsimple \
          -Wl,-R../sharedLib -o ver3PeerApp
```

It will be a nice little exercise for the reader to verify that in all other aspects (versioning information, symbols presence, runtime outcomes) the example meets his/her expectations.

Version Script Syntax Overview

The version scripts shown in the code examples so far feature only a subset of the broad range of supported syntax features. The purpose of this section is to provide a brief overview of the supported options.

Version Node

The basic entity of the version script is the version node, the named construct encapsulated between the curly brackets describing certain version, such as

```
LIBXYZ_1.0.6 {

    ... <some descriptors reside here>

};
```

The version node usually encapsulates several keywords controlling different aspects of the versioning process, whose variety will be discussed in more detail shortly.

Version Node Naming Rules

The node name is typically chosen to precisely describe the full version described by the node. Usually, the name ends with digits separated by dots or underscores. It is a common sense practice that the nodes representing the later versions come after the nodes representing earlier versions.

However, this is just a practice that makes the life of humans easier. The linker does not particularly care how you name your version nodes, nor does it care what order they appear in the file. All that it really requires that the names are different.

A similar situation is with the dynamic libraries and their client binaries. What really matters to them is the **chronology** in which the version nodes were added to the version files—which particular version was present at the time they were built.

Symbols Export Control

The global and local modifiers of a version node directly control the symbol exporting. The semicolon-separated list of symbols declared under the global label will be exported, as opposed to the symbols declared under the local label.

```
LIBXYZ_1.0.6 {
    global:
        first_function; second_function;
    local:
        *;
};
```

Even though it is not the primary topic of the versioning scheme, this mechanism of exporting the symbols is in fact completely legitimate (and in many aspects the most elegant) way of specifying the list of exported symbols. An example of how this mechanism works will be provided in subsequent sections.

Wildcard Support

The version script supports the same set of wildcards that the shells support for expression matching operations. For example, the following version script declares as global all the functions whose name starts with "first" or "second:"

```
LIBXYZ_1.0.6 {
    global:
        first*; second*;
    local:
        *;
};
```

Additionally, the asterisk under the local label specifies all other functions to be of the local scope (those not to be exported). Filenames specified under double quotes are to be taken verbatim, regardless of any wildcard characters they may contain.

Linkage Specifier Support

The version script may be used to specify the extern "C" (no name mangling) or extern "C++" linkage specifier.

```
LIBXYZ_1.0.6 {
    global:
        extern "C" {
            first_function;
        }
    local:
        *;
};
```

Namespace Support

The version scripts also support the use of namespace in specifying the affiliation of the versioned and/or exported symbols.

```
LIBXYZ_1.0.6 {
    global:
        extern "C++" {
            libxyz_namespace::*
        }
    local:
        *;
};
```

Unnamed Node

An unnamed node can be used to specify the unversioned symbols. Additionally, its purpose may be to host the symbols export specifiers (global and/or local).

In fact, when the control over the symbol export is your only motive for using the versioning script mechanism, it is very usual to have a version script containing only one unnamed node.

Version Script Side Feature: Symbol Visibility Control

A side feature of the version script mechanism is that it also provides control over the symbol visibility. The symbols listed in the version script node under the *global* tab end up being exported, whereas the symbols listed under the *local* tab do not get exported.

It is perfectly legal to use the version script mechanism solely for the purpose of specifying the symbols to export. However, it is highly recommended in such case to use the unnamed script version nodes, as demonstrated in the simple demo illustrated by Figure 10-26.

```
milan@milan$ ls -alg
total 20
drwxrwxr-x 2 milan 4096 .
drwxrwxr-x 4 milan 4096 ..
-rwxrwxr-x 1 milan  170 build.sh
-rw-rw-r-- 1 milan   53 exportControlScript
-rw-rw-r-- 1 milan  169 scriptVisibilityControl.c
milan@milan$ cat build.sh
gcc -Wall -fPIC -c scriptVisibilityControl.c
gcc -shared scriptVisibilityControl.o \
    -Wl,--version-script,exportControlScript \
    -o libscriptcontrolsexportdemo.so
milan@milan$ cat scriptVisibilityControl.c
int first_function(int x)
{
        return 0;
}

int second_function(int x)
{
        return 0;
}

int third_function(int x)
{
        return 0;
}
milan@milan$ cat exportControlScript
{
        global:
                first_function;
        local:
                *;
};

milan@milan$ ./build.sh
milan@milan$ nm libscriptcontrolsexportdemo.so | grep function
0000037c T first_function
00000386 t second_function
00000390 t third_function
milan@milan$
```

Figure 10-26. *Version script can be used as the most elegant way of controlling the symbol visibility, as it does not require any modifications of the source code*

Windows Dynamic Libraries Versioning

The versioning implementation in Windows follows an identical set of principles as its Linux counterpart. The code changes that significantly break away from the existing runtime functionality or require rebuilding of client binaries lead to the major version changes. The additions/expansions of the provided functionality that do not disrupt the functionality of the existing client binaries qualify for the minor version changes.

The code changes that mostly affect the details of inner functionality are referred to in Linux as **patches** and in Windows as **build versions**. Other than the obvious naming differences, there are no substantial differences between the two concepts.

DLL Version Information

As with Linux dynamic libraries, the versioning information of Windows dynamic libraries (DLL) is optional. Unless a conscious design effort is made to specify such information, it will not be appearing as part of the DLL. As a good design rule, however, all the major DLL vendors (starting with Microsoft, of course) make sure the dynamic libraries they supply carry the version information. When available, the DLL version information is provided as a dedicated tab on the file property pages, which may be retrieved by right-clicking the file icon in the File Explorer pane, as shown in Figure 10-27.

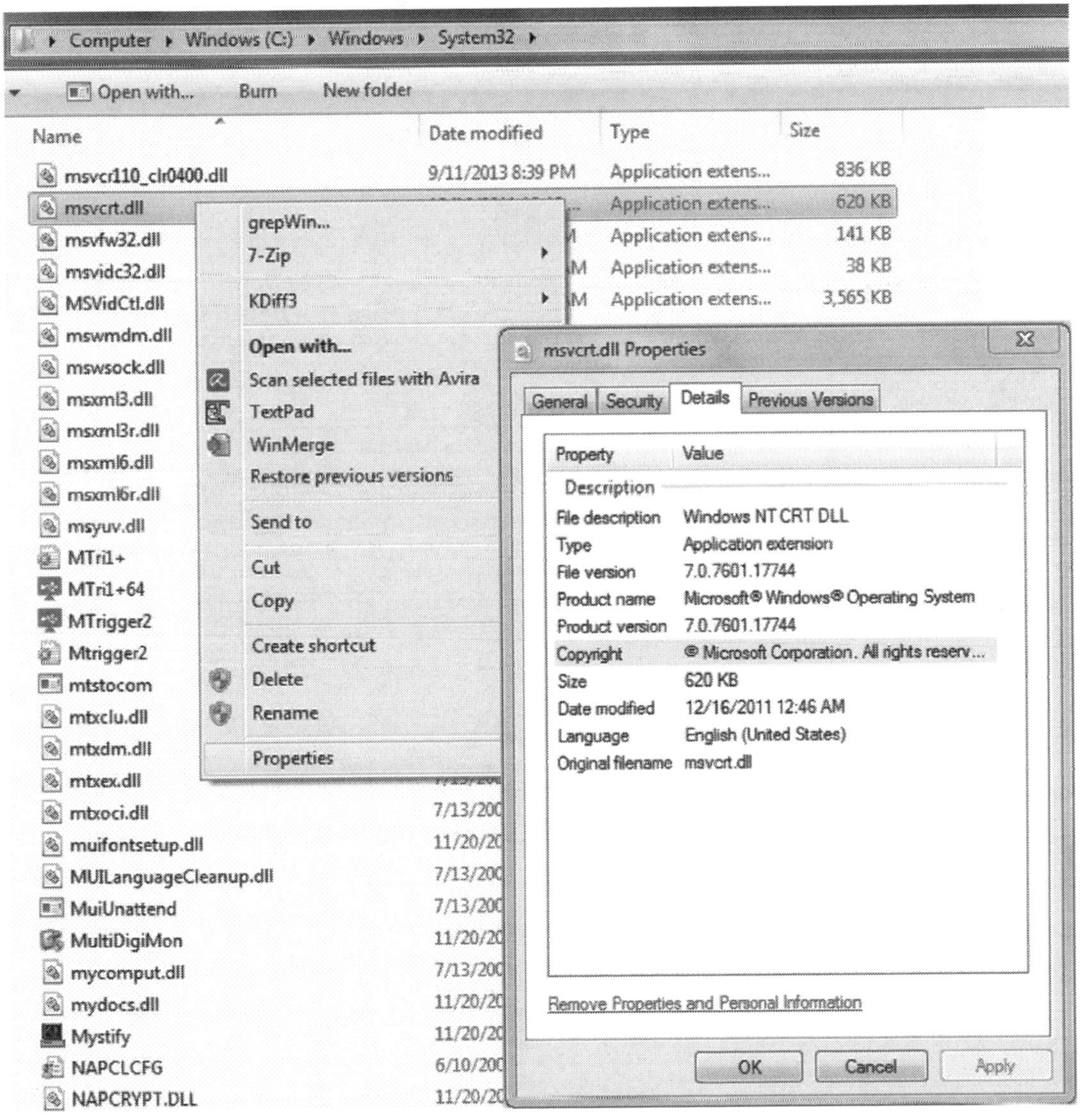

Figure 10-27. Example of DLL version information

Specifying DLL Version Information

In order to illustrate the most important aspects of the Windows DLL versioning, a demo Visual Studio solution is created featuring two projects:

- VersionedDLL project, which builds the DLL whose version information is provided

- VersionedDLLClientApp project, which builds the client application that loads the versioned DLL and tries to retrieve its versioning information

The usual way to supply the versioning information to the DLL project is to add the dedicated version resource element to the library resource file, as shown in Figure 10-28.

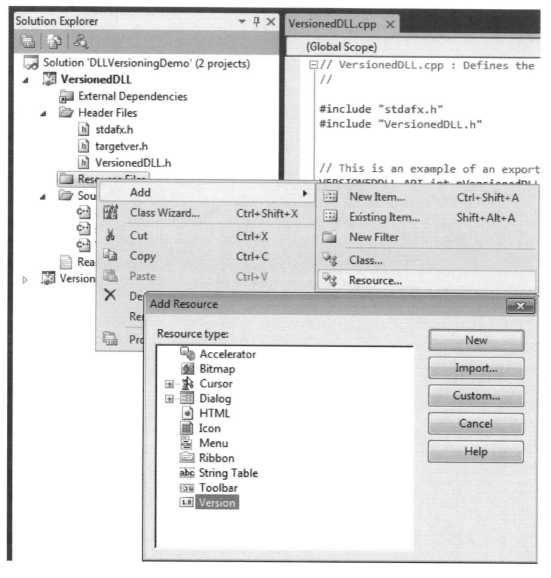

Figure 10-28. Adding the version field to the project resource file

Once the version resource is added to the DLL project resource file, it can be viewed and modified through the Visual Studio resource editor.

As Figure 10-29 indicates, the versioning information provides two distinct versioning components, FILEVERSION as well as PRODUCTVERSION. Despite the fact that in the vast majority of real life scenarios these two components have the identical values, certain differences exist in how the values of these components are set. If a DLL is used in more than one project, its file version number will likely be noticeably larger that its product version number.

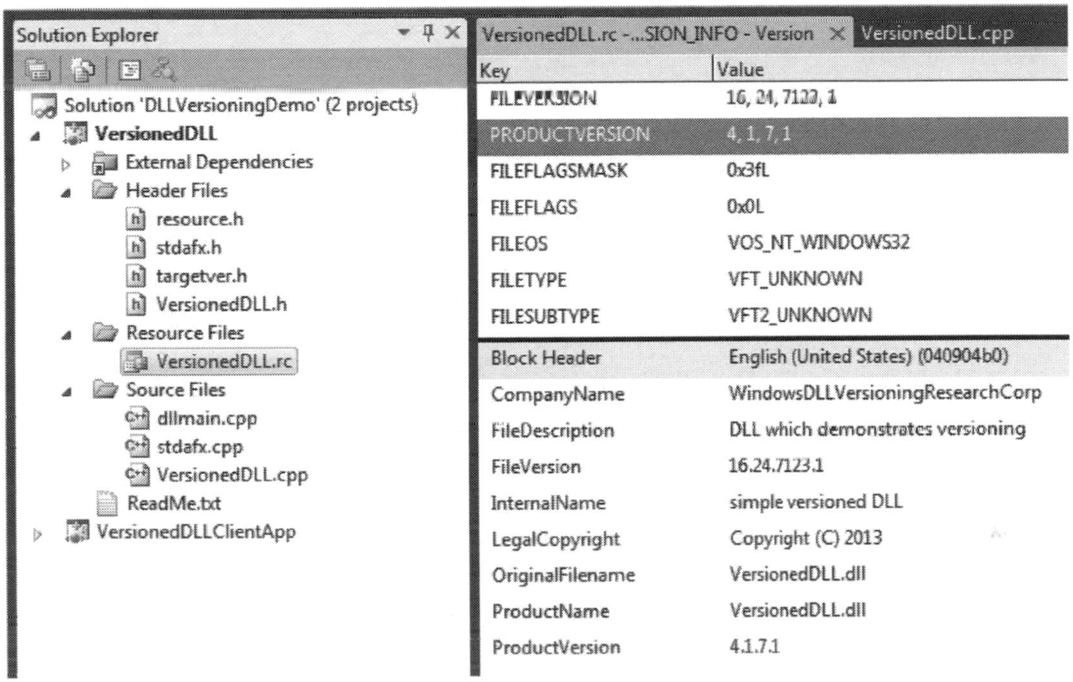

Figure 10-29. *Using the Visual Studio editor to set file version and product version information*

Normally, when the DLL is just created, the version (major version, minor version, build number) are typically set to fairly small and fairly rounded values, such as 1.0.0. In this example, however, I've purposefully chosen for the sake of a convincing demo to not only set the versioning information to fairly large numeric values, but also so that the FILEVERSION values differ from the PRODUCTVERSION values.

When library is built, the version information specified by editing the version resource file can be viewed by right-clicking the file icon on the File Explorer pane, and choosing the Properties menu item (Figure 10-30).

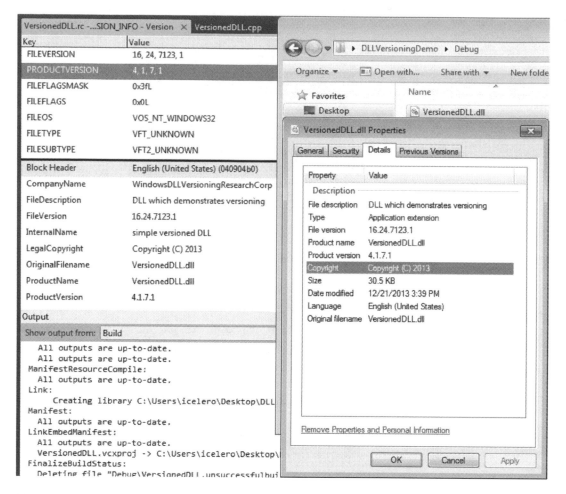

Figure 10-30. Set values appearing in the properties of built DLL binary file

Querying and Retrieving DLL Version Information

The DLL version information may be of particular importance to several interested parties and in a multitude of scenarios. The client binaries whose functionality critically depends on the DLL version may want to programmatically examine the DLL version details in order to take an appropriate course of further actions. The installation/deployment packages may first retrieve the version information of the existing DLLs in order to decide whether or not to replace/overwrite the existing DLLs with newer versions of the same file. Finally, the persons performing the system administration maintenance or troubleshooting duties may wish to take a closer look at the DLL version.

In this section I will focus primarily on the programmatic ways the DLL version information may be retrieved.

VERSIONINFO Structure

The DLLVERSIONINFO structure, declared in the `<shlwapi.h>` header file, is typically used to pass the versioning information. Figure 10-31 shows its layout details.

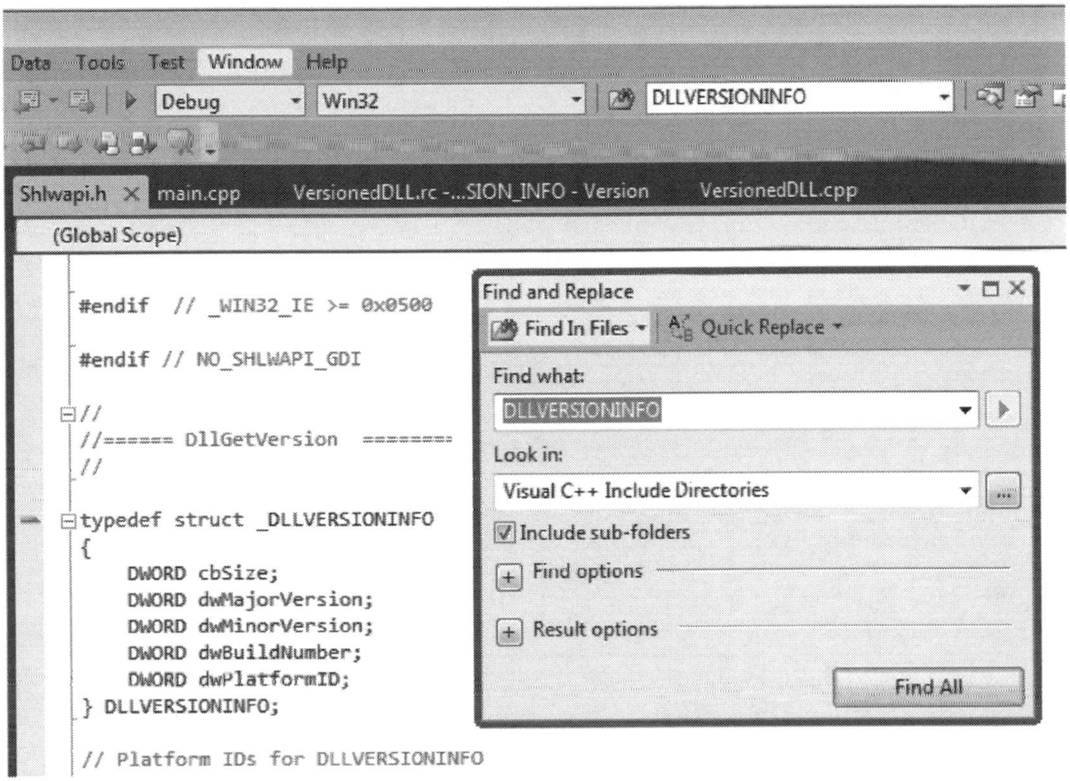

Figure 10-31. DLLVERSIONINFO structure

Linking Requirements

The software modules that need to access the versioning related functionality must be linked with the `version.dll` (i.e., its import library `version.lib` must be specified in the list of linker inputs), as illustrated by Figure 10-32.

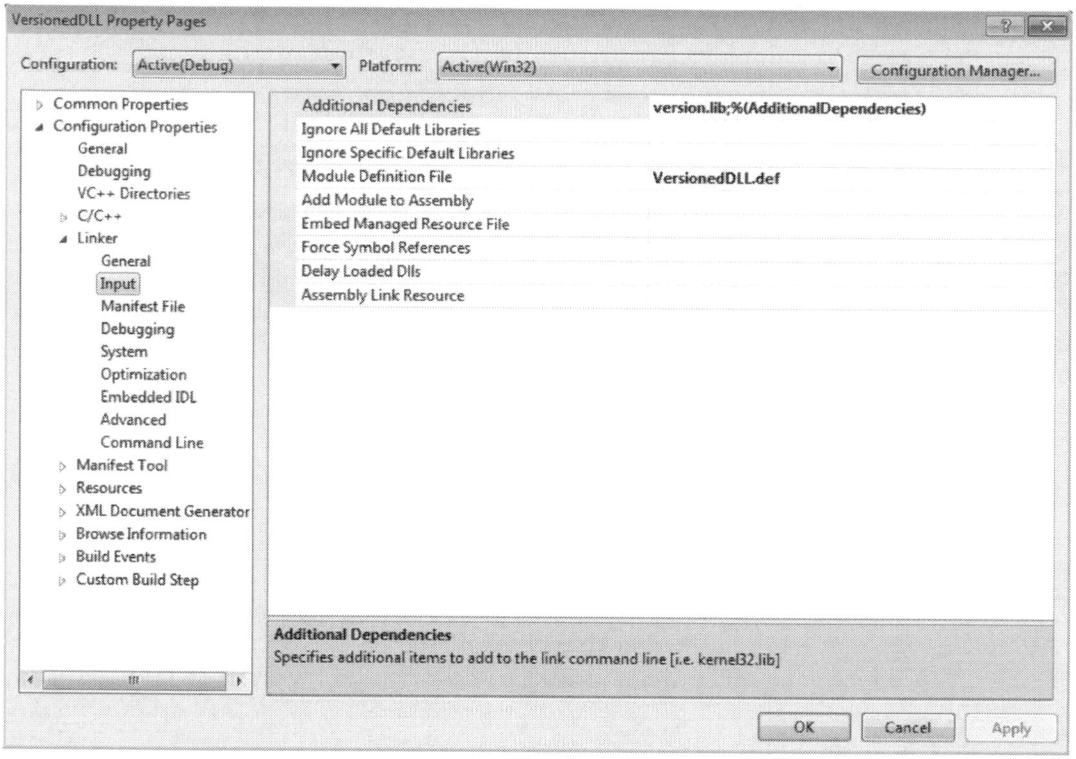

Figure 10-32. *Linking against version.lib (version.dll) is required*

The ways that the DLL version information may be retrieved will be discussed next.

Elegant Way: Calling the DLL's DllGetVersion Function

Well-designed DLLs typically export the implementation of the DllGetVersion() function, whose signature adheres to the following specification:

```
HRESULT CALLBACK DllGetVersion( DLLVERSIONINFO *pdvi);
```

This is mentioned in the in MSDN documentation at http://msdn.microsoft.com/enus/library/windows/desktop/bb776404(v=vs.85).aspx. The DLLs provided by the Microsoft typically provide the expected functionality.

It is not complicated for the custom designed DLLs to implement it as well. Here is the outline of the recipe: the function prototype must be properly declared and exported, illustrated by Listing 10-20, as well as by Figure 10-33.

Listing 10-20. VersionedDll.h

```
// The following ifdef block is the standard way of creating macros which make exporting
// from a DLL simpler. All files within this DLL are compiled with the VERSIONEDDLL_EXPORTS
// symbol defined on the command line. This symbol should not be defined on any project
// that uses this DLL. This way any other project whose source files include this file see
// VERSIONEDDLL_API functions as being imported from a DLL, whereas this DLL sees symbols
// defined with this macro as being exported.
```

```
#ifdef VERSIONEDDLL_EXPORTS
#define VERSIONEDDLL_API __declspec(dllexport)
#else
#define VERSIONEDDLL_API __declspec(dllimport)
#endif

#include <Shlwapi.h>

VERSIONEDDLL_API HRESULT CALLBACK DllGetVersion(DLLVERSIONINFO* pdvi);
```

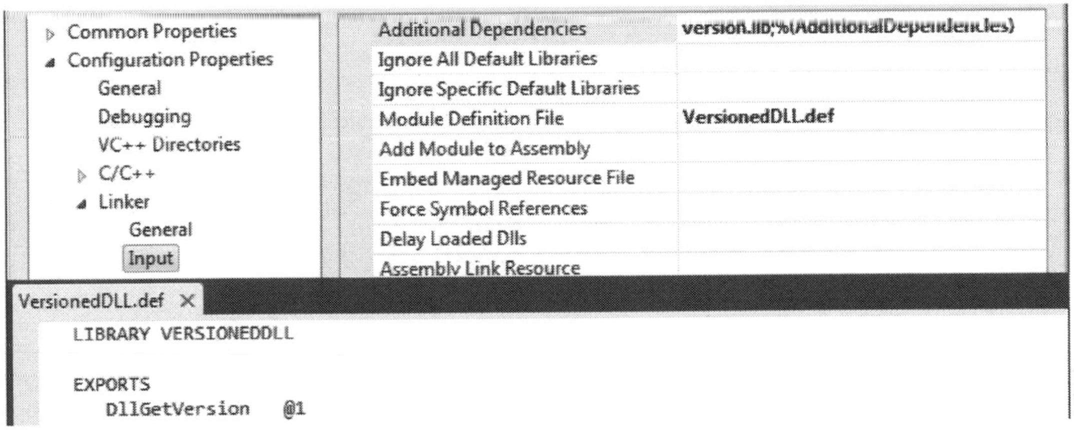

Figure 10-33. *Properly exporting DllGetVersion() function from DLL*

There are several ways the function may be implemented.

- The DLLVERSIONINFO structure members may be set to the predetermined set of values. It is preferred to have the version values in the form of parameterized constants (instead of literal constants).

- The DLLVERSIONINFO structure may be populated by loading the DLL resources, extracting the version information strings, and parsing out the details about major, minor, build version.

Listing 10-21 illustrates the combination of both methods. If the version resource retrieval failed, the predetermined values may be returned. (For the sake of simplicity, the literal constants are used in this listing. We all know that it can be accomplished in more structured way).

Listing 10-21. VersionedDLL.cpp

```
#define SERVICE_PACK_HOTFIX_NUMBER  (16385)

VERSIONEDDLL_API HRESULT CALLBACK DllGetVersion(DLLVERSIONINFO* pdvi)
{
        if(pdvi->cbSize != sizeof(DLLVERSIONINFO) &&
           pdvi->cbSize != sizeof(DLLVERSIONINFO2))
        {
                return E_INVALIDARG;
        }
```

```
        if(FALSE == extractVersionInfoFromThisDLLResources(pdvi))
        {
                // should not happen that we end up here,
                // but just in case - try to save the day
                // by sticking in the actual version numbers
                // TBD: use parametrized value instead of literals
                pdvi->dwMajorVersion = 4;
                pdvi->dwMinorVersion = 1;
                pdvi->dwBuildNumber  = 7;
                pdvi->dwPlatformID   = DLLVER_PLATFORM_WINDOWS;
        }
        if(pdvi->cbSize == sizeof(DLLVERSIONINFO2))
        {
                DLLVERSIONINFO2 *pdvi2 = (DLLVERSIONINFO2*)pdvi;
                pdvi2->dwFlags = 0;
                pdvi2->ullVersion = MAKEDLLVERULL(pdvi->dwMajorVersion,
                                                  pdvi->dwMinorVersion,
                                                  pdvi->dwBuildNumber,
                                                  SERVICE_PACK_HOTFIX_NUMBER);
        }
        return S_OK;
}
```

The details of the function that extracts the version information from the DLL resources follow immediately in Listing 10-22.

Listing 10-22. VersionedDLL.cpp (upper part)

```
extern HMODULE g_hModule;

BOOL extractVersionInfoFromThisDLLResources(DLLVERSIONINFO* pDLLVersionInfo)
{
    static WCHAR fileVersion[256];
    LPWSTR lpwstrVersion = NULL;
       UINT  nVersionLen  = 0;
    DWORD  dwLanguageID  = 0;
    BOOL   retVal;

    if(NULL == pDLLVersionInfo)
        return FALSE;

    HRSRC hVersion = FindResource(g_hModule,
                                  MAKEINTRESOURCE(VS_VERSION_INFO),
                                  RT_VERSION );
    if(NULL == hVersion)
        return FALSE;

    HGLOBAL hGlobal = LoadResource( g_hModule, hVersion );
    if(NULL == hGlobal)
        return FALSE;
```

```
    LPVOID lpstrFileVersionInfo  = LockResource(hGlobal);
    if(NULL == lpstrFileVersionInfo)
        return FALSE;

    wsprintf(fileVersion, L"\\VarFileInfo\\Translation");
    retVal = VerQueryValue ( lpstrFileVersionInfo,
                             fileVersion, (LPVOID*)&lpwstrVersion, (UINT *)&nVersionLen);
    if(retVal && (4 == nVersionLen))
    {
        memcpy(&dwLanguageID, lpwstrVersion, nVersionLen);
        wsprintf(fileVersion, L"\\StringFileInfo\\%02X%02X%02X%02X\\ProductVersion",
                          (dwLanguageID & 0xff00)>>8,
                           dwLanguageID & 0xff,
                          (dwLanguageID & 0xff000000)>>24,
                          (dwLanguageID & 0xff0000)>>16);
    }
    else
    wsprintf(fileVersion,L"\\StringFileInfo\\%04X04B0\\ProductVersion",GetUserDefaultLangID());

    if(FALSE == VerQueryValue (lpstrFileVersionInfo,
                               fileVersion,
                               (LPVOID*)&lpwstrVersion,
                               (UINT *)&nVersionLen))
    {
        return FALSE;
    }

    LPWSTR pwstrSubstring = NULL;
    WCHAR* pContext = NULL;
    pwstrSubstring = wcstok_s(lpwstrVersion, L".", &pContext);
    pDLLVersionInfo->dwMajorVersion = _wtoi(pwstrSubstring);

    pwstrSubstring = wcstok_s(NULL, L".", &pContext);
    pDLLVersionInfo->dwMinorVersion = _wtoi(pwstrSubstring);

    pwstrSubstring = wcstok_s(NULL, L".", &pContext);
    pDLLVersionInfo->dwBuildNumber = _wtoi(pwstrSubstring);

    pwstrSubstring = wcstok_s(NULL, L".", &pContext);
    pDLLVersionInfo->dwPlatformID = _wtoi(pwstrSubstring);

    pDLLVersionInfo->cbSize = 5*sizeof(DWORD);

    UnlockResource( hGlobal );
    FreeResource( hGlobal );

    return TRUE;
}
```

The important part of the recipe is that the good moment to capture the value of DLL's module handle is when the DllMain() function gets called, as shown in Listing 10-23.

Listing 10-23. dllmain.cpp

```
// dllmain.cpp : Defines the entry point for the DLL application.
#include "stdafx.h"

HMODULE g_hModule = NULL;

BOOL APIENTRY DllMain( HMODULE hModule,
                       DWORD   ul_reason_for_call,
                       LPVOID  lpReserved
                                             )
{
        switch (ul_reason_for_call)
        {
        case DLL_PROCESS_DETACH:
        g_hModule = NULL;
                break;
        case DLL_PROCESS_ATTACH:
                g_hModule = hModule;
        case DLL_THREAD_ATTACH:
        case DLL_THREAD_DETACH:
                break;
        }
        return TRUE;
}
```

Finally, Listing 10-24 shows how the client binary retrieves the versioning information.

Listing 10-24. main.cpp (client app)

```
BOOL extractDLLProductVersion(HMODULE hDll, DLLVERSIONINFO* pDLLVersionInfo)
{
    if(NULL == pDLLVersionInfo)
        return FALSE;

    DLLGETVERSIONPROC pDllGetVersion;
    pDllGetVersion = (DLLGETVERSIONPROC) GetProcAddress(hDll, "DllGetVersion");
    if(NULL == pDllGetVersion)
        return FALSE;

    ZeroMemory(pDLLVersionInfo, sizeof(DLLVERSIONINFO));
    pDLLVersionInfo->cbSize = sizeof(DLLVERSIONINFO);
    HRESULT hr = (*pDllGetVersion)(pDLLVersionInfo);
    if(FAILED(hr))
        return FALSE;

    return TRUE;
}
```

Brutal Alternative: Examining File Version Directly

If it happens that the DLL does not export the DllGetVersion() function, you may resort to the more brutal measure of extracting the versioning information embedded in the file resources. The complete effort of implementing this approach resides on the client binary side. As can be easily concluded by comparing the following code with the code laid out in the previous approach description, the same methodology is applied, based on loading the resources from the file, extracting the version string and extracting the version numbers thereof (see Listing 10-25).

Listing 10-25. main.cpp (client app)

```
BOOL versionInfoFromFileVersionInfoString(LPSTR lpstrFileVersionInfo,
                                          DLLVERSIONINFO* pDLLVersionInfo)
{
    static WCHAR fileVersion[256];
    LPWSTR lpwstrVersion    = NULL;
    UINT   nVersionLen  = 0;
    DWORD  dwLanguageID = 0;
    BOOL   retVal;

    if(NULL == pDLLVersionInfo)
        return FALSE;

    wsprintf(fileVersion, L"\\VarFileInfo\\Translation");
    retVal = VerQueryValue ( lpstrFileVersionInfo,
                             fileVersion, (LPVOID*)&lpwstrVersion, (UINT *)&nVersionLen);
    if(retVal && (4 == nVersionLen))
    {
        memcpy(&dwLanguageID, lpwstrVersion, nVersionLen);
        wsprintf(fileVersion, L"\\StringFileInfo\\%02X%02X%02X%02X\\FileVersion",
                (dwLanguageID & 0xff00)>>8,
                 dwLanguageID & 0xff,
                (dwLanguageID & 0xff000000)>>24,
                (dwLanguageID & 0xff0000)>>16);
    }
    else
        wsprintf(fileVersion,L"\\StringFileInfo\\%04X04B0\\FileVersion",GetUserDefaultLangID());

    if(FALSE == VerQueryValue (lpstrFileVersionInfo,
                               fileVersion,
                               (LPVOID*)&lpwstrVersion,
                               (UINT *)&nVersionLen))
    {
        return FALSE;
    }

    LPWSTR pwstrSubstring = NULL;
    WCHAR* pContext = NULL;
    pwstrSubstring = wcstok_s(lpwstrVersion, L".", &pContext);
    pDLLVersionInfo->dwMajorVersion = _wtoi(pwstrSubstring);
```

```
        pwstrSubstring = wcstok_s(NULL, L".", &pContext);
        pDLLVersionInfo->dwMinorVersion = _wtoi(pwstrSubstring);

        pwstrSubstring = wcstok_s(NULL, L".", &pContext);
        pDLLVersionInfo->dwBuildNumber = _wtoi(pwstrSubstring);

        pwstrSubstring = wcstok_s(NULL, L".", &pContext);
        pDLLVersionInfo->dwPlatformID = _wtoi(pwstrSubstring);

        pDLLVersionInfo->cbSize = 5*sizeof(DWORD);
        return TRUE;
}

BOOL extractDLLFileVersion(DLLVERSIONINFO* pDLLVersionInfo)
{
        DWORD dwVersionHandle = 0;
        DWORD dwVersionInfoSize = GetFileVersionInfoSize (DLL_FILENAME, &dwVersionHandle);
        if(0 == dwVersionInfoSize)
            return FALSE;

        LPSTR lpstrFileVersionInfo = (LPSTR) malloc (dwVersionInfoSize);
        if (lpstrFileVersionInfo == NULL)
            return FALSE;

        BOOL bRetValue = GetFileVersionInfo(DLL_FILENAME,
                                            dwVersionHandle,
                                            dwVersionInfoSize,
                                            lpstrFileVersionInfo);
        if(bRetValue)
        {
            bRetValue = versionInfoFromFileVersionInfoString(lpstrFileVersionInfo, pDLLVersionInfo);
        }

        free (lpstrFileVersionInfo);
        return bRetValue;
}

int main(int argc, char* argv[])
{
        //
        // Examining the DLL file ourselves
        //
        memset(&dvi, 0, sizeof(DLLVERSIONINFO));
        if(extractDLLFileVersion(&dvi))
        {
            printf("DLL File Version (major, minor, build, platformID) = %d.%d.%d.%d\n",
                    dvi.dwMajorVersion, dvi.dwMinorVersion,
                    dvi.dwBuildNumber, dvi.dwPlatformID);
        }
```

```
    else
        printf("DLL File Version extraction failed\n");

    FreeLibrary(hDll);

    return 0;
}
```

Finally, the result of running the demo app which demonstrates both approaches (DLL querying and "brute force") is shown in Figure 10-34.

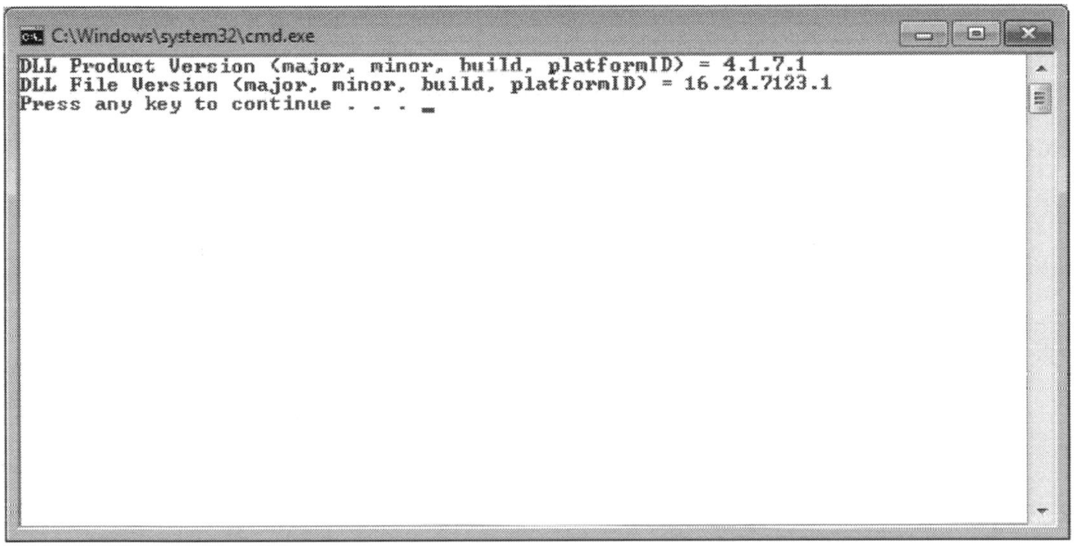

Figure 10-34. *Programmatically extracting DLL product version as well as the file version*

CHAPTER 11

■ ■ ■

Dynamic Libraries: Miscellaneous Topics

After understanding the most profound ideas behind the concept of dynamic libraries, and before diving into the details of the toolbox of the software professional dealing with libraries on daily basis, it is a good moment to take a closer look at a few remaining issues. First, let's take a closer look at the concept of plug-ins, the omnipresent mechanism of seamlessly extending the basic functionality of a framework. Then, I will point out a few practical implications stemming from the concept of dynamic libraries. Finally, I will take a closer look at a few miscellaneous topics which a developer may encounter in the everyday work.

Plug-in Concept

Probably the most important concept made possible by the advancement of dynamic linking is the concept of plug-ins. There is nothing substantially hard to understand in the concept itself, as we encounter it in plethora of everyday scenarios, most of which don't require any technical background. A good example of the plug-in concept is the drill and the variety of drill bits that can be changed as per the needs of particular situation and the decision of the end user (Figure 11-1).

Figure 11-1. Drill and bits, an everyday example of the plug-in concept

The software concept of plug-ins follows the same principle. Basically, there is a major application (or execution environment) that performs a certain action on a certain processing subject (for example, the photo processing application that modifies a picture's properties), and there is a set of modules that specialize in performing a very specific action on the processing subject (for example, blurring filter, sharpening filter, sepia filter, color contrast filter, high pass filter, averaging filter, etc.), a concept which is very easy to comprehend.

But that's not all.

Not all of the systems comprising the flagship application and the associated modules deserve to be called "plug-in architecture." In order for architecture to support the plug-in model, the following requirements need to be satisfied as well:

- Adding or removing the plug-ins should not require that the application be recompiled; instead, the application should be capable of determining the availability of plug-ins at runtime.

- The modules should export their functionality through a runtime loadable mechanism of some kind.

- The system should be functional regardless of which plug-ins are available at runtime to end user.

In reality, the above requirements are typically supported through the following design decisions:

- Plug-ins arc implemented as dynamic libraries. Regardless of the inner functionality, all plug-in dynamic libraries export the standardized interface (a set of functions allowing the application to control the plug-in execution).

- The application loads the plug-ins through the process of dynamic library loading. The following two options are typically supported:

    - The application looks at the predefined folder and tries loading all the dynamic libraries that it finds there at runtime. Upon loading, it tries to find the symbols corresponding to the interface that the plug-ins are expected to export. If the symbols are not found (or only some of them are found), the plug-in library is unloaded.

    - The user, through a dedicated GUI option at runtime, specifies the plug-in location and tells the application to load the plug-in and start providing its functionality.

Rules of Exporting

Rigorously strict rules for each and every plug-in architecture do not exist. However, a common sense set of guidelines does exist. According to the paragraph explaining the impact of the C++ language on linker problems, the majority of plug-in architectures tend to follow the simplest possible scheme in which a plug-in exports a pointer to the interface comprised of C-linkage functions.

Even though the plug-ins' inner functionality may be implemented as a C++ class, such a class typically implements the interface exported by its dynamic library container, and passing a pointer to the class instance (casted as the pointer to the interface) to the application is usual practice.

Popular Plug-in Architectures

There is vast variety of popular programs that support the plug-in architecture, such as (but not limited to):

- Image processing applications (Adobe Photoshop, etc.)

- Video processing applications (Sony Vegas, etc.)

- Sound processing (Steinberg VST plug-in architecture, universally supported across all major audio editors)

- Multimedia frameworks (GStreamer, avisynth) and popular applications (Winamp, mplayer)

- Text editors (a vast number of which have plug-ins providing certain functionalities)

- Software development integrated development environments (IDEs) supporting a variety of features through the plug-ins

- Version control systems' front-end GUI applications

- Web browsers (NPAPI plug-in architecture)

- Etc.

For each of these plug-in architectures there is typically a published plug-in interface document stipulating in detail the interaction between the application and the plug-in.

Tips and Tricks

The very last step on your journey to fully understanding the concept of dynamic libraries requires that you take a little step back and organize everything you've learned so far into another set of simple facts. Formulating things in a different manner may sometimes mean a lot of difference in the domain of everyday design practices.

Practical Implications of Working with Dynamic Libraries

After all the details about the dynamic libraries have been examined, the most potent truth about them is that linking against dynamic libraries is kind of linking on promise. Indeed, during the build stage, all that the client executable worries about are the dynamic libraries symbols. It is only at the runtime loading stage that the contents of the dynamic library sections (code, data, etc.) come to play. There are several real-life implications stemming from the described set of circumstances.

Compartmentalized, Faster Development

The concept of dynamic libraries allows a lot of freedom for a programmer. As long as the set of symbols important to the client executable does not change, the programmer is free to keep modifying the actual dynamic library code as long and as much as desired.

This simple fact has a tremendous impact on the programming everyday routines, as it tends to greatly reduce unnecessary compile time. Instead of having to recompile the whole body of code whenever a minuscule change of code happens, by using the dynamic libraries, the programmers can reduce the need to rebuild the code to the dynamic library itself. It is no wonder that programmers often decide to host code under development in the dynamic library, at least up until the development is completed.

Runtime Quick Substitution Ability

At build time, the client binary doesn't need the fully fledged dynamic library with all bells and whistles in place. Instead, all that the client binary really needs at build time is the set of dynamic library's symbols—nothing more and nothing less than that.

This is really interesting. Please take a deep breath, and let's take a look at what this claim really means.

The dynamic library binary file that you use at build time and the dynamic library file that is loaded at runtime may be substantially different in every aspect, except in one: the symbols must match.

In other words (and yes, this is true and exactly how it was meant to be), for the statically-aware building purposes you may use dynamic library whose code (flash and blood) is yet about to be implemented, but its symbols (the skeleton) are already in their final shape.

Or, you may use a library whose code you know will change, as long as you may be assured that the set of exported symbols will not change.

Or, you may use the dynamic library suitable for one specific flavor (such as language pack) at build time, but link at runtime with another dynamic library—as long as the both dynamic library binaries export the same set of symbols.

This is really, really interesting. An extreme example of how we may benefit from this important finding happens in the domain of Android native programming. During the effort to develop a module (dynamic library or native application), it is not completely uncommon for whole teams of developers unnecessarily and unwisely to take the time-consuming path of adding their source code into the gigantic Android source tree whose building may take several hours.

Alternatively, a far more effective procedure is to develop a module as a standalone Android project, unrelated to the Android source tree. In a matter of minutes, the Android native dynamic libraries necessary to complete the build phase may be copied ("adb pulled" in Android lingo) from any working Android device/phone and added to the project build structure. Instead taking several hours, the build procedure now takes several minutes at most.

Even though the code (the .text section) of the dynamic library pulled from the nearest available Android phone may significantly differ from the code found in the Android source tree, the list of symbols is very likely identical in both dynamic libraries. Obviously, the quick replacement library pulled from the Android device may satisfy the build requirements, whereas at runtime "the right one" of the dynamic library binary will be loaded.

Miscellaneous Tips

In the remainder of this chapter, I will cover the following interesting tidbits of knowledge:

- Converting the dynamic library to executable
- Conflicting runtime memory handling scenarios of Windows libraries
- Linker weak symbols

Converting Dynamic Library to Executable

As was pointed out earlier in the introductory discussions about dynamic libraries, the difference between the dynamic library and executable is in the fact that later has startup routines which allow the kernel to actually start the execution. In all other aspects, especially if compared to the static libraries, it appears that the dynamic library and executable are of the same nature, such as the binary file in which all the references have been resolved.

Given so many similarities and so little differences, is it possible to convert the dynamic library to the executable?

The answer to this question is positive. It's most certainly possible on Linux (I'm still looking to confirm the claim on Windows as well). As a matter of fact, the library implementing C runtime library (libc.so) is in fact truly executable. When invoked by typing its filename in the shell window, you get the response shown in Figure 11-2.

```
milan@milan$ /lib/i386-linux-gnu/libc.so.6
GNU C Library (Ubuntu EGLIBC 2.15-0ubuntu10) stable release version 2.15, by Roland
McGrath et al.
Copyright (C) 2012 Free Software Foundation, Inc.
This is free software; see the source for copying conditions.
There is NO warranty; not even for MERCHANTABILITY or FITNESS FOR A
PARTICULAR PURPOSE.
Compiled by GNU CC version 4.6.3.
Compiled on a Linux 3.2.14 system on 2012-04-19.
Available extensions:
        crypt add-on version 2.1 by Michael Glad and others
        GNU Libidn by Simon Josefsson
        Native POSIX Threads Library by Ulrich Drepper et al
        BIND-8.2.3-T5B
libc ABIs: UNIQUE IFUNC
For bug reporting instructions, please see:
<http://www.debian.org/Bugs/>.
milan@milan$
```

Figure 11-2. *Running libc.so as executable file*

The question that naturally comes next is how to implement the library in order to make it executable? The following recipe makes it possible:

- Implement the main function inside the dynamic library—the function whose prototype is

  ```
  int main(int argc, char* argv[];
  ```

- Declare the standard main() function as the library entry point. Passing the -e linker flag is how you accomplish this task.

  ```
  gcc -shared -Wl,-e,main -o<libname>
  ```

- Turn the main() function into a no-return function. This can be done by inserting the _exit(0) call as the last line of the main() function.

- Specify the interpreter to be the dynamic linker. The following line of code would do it:

  ```
  #ifdef __LP64__
  const char service_interp[] __attribute__((section(".interp"))) =
          "/lib/x86_64-linux-gnu/ld-linux-x86-64.so.2";
  #else
  const char service_interp[] __attribute__((section(".interp"))) =
          "/lib/ld-linux.so.2";
  #endif
  ```

- Built the library without the optimization (with the -O0 compiler flag).

A simple demo project is made to illustrate the idea. In order to prove the truly dual nature of the dynamic library (i.e., even though it now can be run as executable, it still remains capable of functioning as a regular dynamic library), the demo project contains not only the demo dynamic library, but also the executable that dynamically loads it and calls its printMessage() function. The Listing 11-1 illustrates the details of executable shared library project:

Listing 11-1.

file: executableSharedLib.c

```
#include "sharedLibExports.h"
#include <unistd.h> // needed for the _exit() function

// Must define the interpretor to be the dynamic linker
#ifdef __LP64__
const char service_interp[] __attribute__((section(".interp"))) =
    "/lib/x86_64-linux-gnu/ld-linux-x86-64.so.2";
#else
const char service_interp[] __attribute__((section(".interp"))) =
    "/lib/ld-linux.so.2";
#endif

void printMessage(void)
{
    printf("Running the function exported from the shared library\n");
}
```

```c
int main(int argc, char* argv[])
{
    printf("Shared library %s() function\n", __FUNCTION__);

    // must make the entry point function to be a 'no-return' function type
    _exit(0);
}
```

file: build.sh

```
g++  Wall  OO  fPIC -I /exports/ -c src/executableSharedLib.c -o src/executableSharedLib.o
g++ -shared -Wl,-e,main ./src/executableSharedLib.o -pthread -lm -ldl -o
../deploy/libexecutablesharedlib.so
```

The Listing 11-2 illustrates the details of the demo app whose purpose is to prove that by becoming executable our shared library did not loose its original functionality:

Listing 11-2.

file: main.c

```c
#include <stdio.h>
#include "sharedLibExports.h"

int main(int argc, char* argv[])
{
    printMessage();
    return 0;
}
```

file: build.sh

```
g++ -Wall -O2 -I../sharedLib/exports/ -c src/main.c -o src/main.o
    g++  ./src/main.o -lpthread -lm -ldl -L../deploy -lexecutablesharedlib -Wl,-Bdynamic -Wl,-R
../deploy -o demoApp
```

When you try to use it, the results shown in Figure 11-3 appear.

```
milan@milan$ tree
.
├── demoApp
├── deploy
│   └── libexecutablesharedlib.so
├── Makefile
├── Notes
│   └── README.txt
├── sharedLib
│   ├── exports
│   │   └── sharedLibExports.h
│   ├── libexecutablesharedlib.so
│   ├── Makefile
│   └── src
│       ├── executableSharedLib.c
│       └── executableSharedLib.o
└── testApp
    ├── demoApp
    ├── Makefile
    └── src
        ├── main.c
        └── main.o

7 directories, 13 files
milan@milan$ ./deploy/libexecutablesharedlib.so
Shared library main() function
milan@milan$ ./demoApp
Running the function exported from the shared library
milan@milan$
```

Figure 11-3. *Illustrating dual nature (dynamic lib, executable) of the demo library*

The project source code tarball provides the more insight into the details.

Conflicting Runtime Memory Handling of Windows Libraries

In general, once the dynamic library gets loaded into the process, it becomes a legitimate part of the process and pretty much inherits all the privileges of the process, including access to the heap (the pool of memory on which dynamic memory allocation runs). For these reasons, it is perfectly normal that a dynamic library function allocates memory buffer, and passes it to a function belonging to the other dynamic library (or to executable code) where the memory can be deallocated when no longer needed.

However, there is a special twist to the whole story which needs to be carefully examined.

Typically, regardless of how many dynamic libraries are loaded into the process, they all link against the same instance of C runtime library, which provides the memory allocation infrastructure—the malloc and free (or in the case of C++, new and delete), as well as the implementation of the list keeping track of the allocated memory buffers. If this infrastructure is unique per process, there are really no reasons why the described scheme in which anybody can deallocate the memory allocated by anybody else should not work.

The interesting case, however, may happen in the domain of Windows programming. Visual Studio provides (at least) two base DLLs on top of which all the executables (applications/dynamic libraries) are built—the usual C runtime library (msvcrt.dll) as well as Microsoft Foundation Classes (MFC) library (mfx42.dll). Sometimes projects requirement may happen to dictate the mixing and matching of the DLLs built upon on different base DLLs, which may immediately cause very unpleasant deviations from the expected rules.

Let's say for the sake of clarity that in the same project you have the following two DLLs loaded at runtime: DLL "A," built on msvcrt.dll, and DLL "B," built on MFC DLL. Let's now assume that DLL "A" allocates memory buffers and passes it to DLL "B," which uses them and then deallocates them. In this case, the attempt to deallocate the memory will result in a crash (the exception looking like the one in Figure 11-4).

```
void __cdecl _unlock (
        int locknum
        )
{
      /*
       * leave the critical section.
       */
      LeaveCriticalSection( _locktable[locknum].lock );
}
```

#ifdef
#pragm
#endif

/***
* _lock
*
*Purpo
*
*
*
*
*

> **Microsoft Visual Studio**
>
> ⚠ Windows has triggered a breakpoint in FLVTranscoderTestApp.exe.
>
> This may be due to a corruption of the heap, which indicates a bug in ▓▓▓▓▓▓▓▓▓▓.exe or any of the DLLs it has loaded.
>
> This may also be due to the user pressing F12 while ▓▓▓▓▓▓▓.exe has focus.
>
> The output window may have more diagnostic information.
>
> [Break]   [Continue]   [Ignore]

Figure 11-4. *Error message dialog typical for between-DLLs-conflict memory issues*

The cause of the problem is that there are two bookkeeping authorities around the available pool of heap memory; both C runtime DLL and MFC DLL maintain their own, separate lists of the allocated buffers (see Figure 11-5).

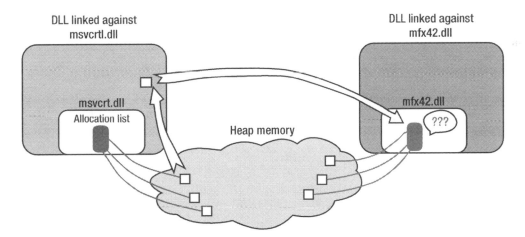

Figure 11-5. *The mechanism of runtime problems caused by unrelated memory allocation bookkeepings maintained by different DLLs*

Normally, when sending the buffer for deallocation, the memory allocation infrastructure searches the list of allocated memory addresses, and if the buffer passed for deallocation is found in the list, the deallocation can be successfully completed. If, however, the allocated buffer is maintained in one list (say, maintained by the C runtime DLL) and passed for deallocation to the other list (say, maintained by the MFC DLL), the buffer's memory address will not be found in the list, and the deallocation call will throw an exception. Even if you handle the exception silently, it is questionable whether the application will be capable of sending the buffer to the right DLL for deallocation, thus causing memory leaks.

For things to be worse, virtually none of the usual memory-bound checking tools have been able to detect and report anything wrong. In the defense of the tools, you can notice that in fact none of the typical memory violations happen in this particular case (such as writing past the buffer boundaries, overwriting buffer address, etc). This all makes the problem unpleasant to deal with, and unless you have an idea upfront about the potential problems, it may be really tough to pinpoint the cause, let alone the solution to the problem.

The solution to the problem is exceptionally simple: the memory buffers allocated in one DLL should be ultimately passed back to the same DLL to be deallocated. The only problem is that in order to apply this simple solution you need to have access to the source code of both DLLs, which may not always be possible.

Linker Weak Symbols Explained

The idea of a linker weak symbol is in its essence similar to the overriding feature of object-oriented languages (which is one of the manifestations of polymorphism principle). When applied to the domain of linking, the idea of weak symbols practically means the following:

- Compilers (most notably, gcc) support the language construct, allowing you to declare a symbol (a function and/or a global or function-static variable) as weak.

The following example demonstrates how to declare a C function as a weak symbol:

```
int __attribute__((weak)) someFunction(int, float);
```

- The linker takes this information to handle such symbol in a very unique way.

  - If another identically named symbol appears during the linking, and is not declared weak, that another symbol will replace the weak symbol.

  - If another identically named symbol appears during the linking and is declared weak, the linker is free to decide which of the two will be actually implemented.

  - The presence of two non-weak (i.e., strong) symbols of the same name is considered an error (the symbol is already defined).

  - If during the linking no other identically named symbols appear, the linker may not implement such symbol. If the symbol is a function pointer, the safeguarding the code is a must (in fact, it is strongly recommended to do it always).

An excellent illustration of the concept of weak symbols is found in Winfred C.H. Lu's blog post at http://winfred-lu.blogspot.com/2009/11/understand-weak-symbols-by-examples.html. The actual real-life scenario of when such features may come handy is described in Andrew Murray's blog post at www.embedded-bits.co.uk/2008/gcc-weak-symbols/.

CHAPTER 12

■ ■ ■

Linux Toolbox

The purpose of this chapter is to introduce the reader to a set of tools (utility programs as well as other methods) for analyzing the contents of the Linux binary files.

Quick Insight Tools

The easiest and the most immediate insight into the nature of a binary file can be obtained by using the file and/or size utility programs.

file Utility Program

The command-line utility named simply file (http://linux.die.net/man/1/file) is used to find out details about just about any file type. It can quickly come in handy because it determines the most basic info about the binary file (Figure 12-1).

```
$ file /usr/bin/gst-inspect-0.10
/usr/bin/gst-inspect-0.10: ELF 32-bit LSB executable, Intel 80386, version 1 (SY
SV), dynamically linked (uses shared libs), for GNU/Linux 2.6.24, BuildID[sha1]=0x41b8f8a4
1450a5b090992220ee852afe2f9d00c2, stripped
$
$
$ file /usr/lib/i386-linux-gnu/xen/libpthread.a
/usr/lib/i386-linux-gnu/xen/libpthread.a: current ar archive
$
$
$ file /lib/i386-linux-gnu/libc-2.15.so
/lib/i386-linux-gnu/libc-2.15.so: ELF 32-bit LSB shared object, Intel 80386, version 1 (SY
SV), dynamically linked (uses shared libs), BuildID[sha1]=0xe4a0e031bf20aaf48f716bee471e36
f5262d7730, for GNU/Linux 2.6.24, stripped
$
```

Figure 12-1. *Using the file utility*

size Utility Program

The command-line utility named size (http://linux.die.net/man/1/size) may be used to instantly obtain an insight into the ELF sections byte lengths (Figure 12-2).

```
$ size /usr/bin/gst-inspect-0.10
   text    data     bss     dec     hex filename
  29056     836      20   29912    74d8 /usr/bin/gst-inspect-0.10
$
$
$ size /lib/i386-linux-gnu/libc-2.15.so
   text    data     bss     dec     hex filename
1696633   11508   11316 1719457  1a3ca1 /lib/i386-linux-gnu/libc-2.15.so
$
$
$ size /usr/lib/i386-linux-gnu/xen/libdl.a
   text    data     bss     dec     hex filename
     83       0       0      83      53 dlopen.o (ex /usr/lib/i386-linux-gnu/xen/libdl.a)
     49       0       0      49      31 dlclose.o (ex /usr/lib/i386-linux-gnu/xen/libdl.a)
     83       0       0      83      53 dlsym.o (ex /usr/lib/i386-linux-gnu/xen/libdl.a)
     91       0       0      91      5b dlvsym.o (ex /usr/lib/i386-linux-gnu/xen/libdl.a)
     49       0       0      49      31 dlerror.o (ex /usr/lib/i386-linux-gnu/xen/libdl.a)
     49       0       0      49      31 dladdr.o (ex /usr/lib/i386-linux-gnu/xen/libdl.a)
     49       0       0      49      31 dladdr1.o (ex /usr/lib/i386-linux-gnu/xen/libdl.a)
     91       0       0      91      5b dlinfo.o (ex /usr/lib/i386-linux-gnu/xen/libdl.a)
     91       0       0      91      5b dlmopen.o (ex /usr/lib/i386-linux-gnu/xen/libdl.a)
```

Figure 12-2. *Using the size utility*

Detailed Analysis Tools

Detailed insight into the binary file properties may be obtained by relying on the collection of utilities collectively referred to as binutils (www.gnu.org/software/binutils/). I will illustrate the use of the ldd, nm, objdump, and readelf utilities. Even though it formally does not belong to the binutils, the shell script called ldd (written by Roland McGrath and Ulrich Drepper) nicely fits in the same compartment of the toolbox, and hence its use will be illustrated as well.

ldd

The command ldd (http://linux.die.net/man/1/ldd) is an exceptionally useful tool, as it shows the complete list of the dynamic libraries which a client binary will try to statically aware load (i.e., the load-time dependencies).

When analyzing the load-time dependencies, ldd first examines the binary file trying to locate the ELF format field in which the list of the most immediate dependencies has been imprinted by the linker (as suggested by the linker command line during the build process).

For each of the dynamic libraries whose names have been found embedded within the client binary file, ldd tries to locate their actual binary files according to the runtime library location search rules (as described in detail in Chapter 7). Once the binaries of the most immediate dependencies have been located, ldd runs the next level of its recursive procedure, trying to find their dependencies. On each of the "second generation" dependencies, ldd runs another round of investigation, and so on.

Once the described recursive search is completed, ldd gathers the list of reported dependencies, trims out the duplicates, and prints out the result (as shown in Figure 12-3).

```
milan@milan:~$ ldd /usr/bin/gst-inspect-0.10
        linux-gate.so.1 =>  (0xb772f000)
        libgstreamer-0.10.so.0 => /usr/lib/i386-linux-gnu/libgstreamer-0.10.so.0 (0xb7633000)
        libgobject-2.0.so.0 => /usr/lib/i386-linux-gnu/libgobject-2.0.so.0 (0xb75e4000)
        libglib-2.0.so.0 => /lib/i386-linux-gnu/libglib-2.0.so.0 (0xb74ea000)
        libpthread.so.0 => /lib/i386-linux-gnu/libpthread.so.0 (0xb74cf000)
        libc.so.6 => /lib/i386-linux-gnu/libc.so.6 (0xb7325000)
        libgmodule-2.0.so.0 => /usr/lib/i386-linux-gnu/libgmodule-2.0.so.0 (0xb7320000)
        libxml2.so.2 => /usr/lib/i386-linux-gnu/libxml2.so.2 (0xb71d3000)
        libm.so.6 => /lib/i386-linux-gnu/libm.so.6 (0xb71a6000)
        librt.so.1 => /lib/i386-linux-gnu/librt.so.1 (0xb719d000)
        libdl.so.2 => /lib/i386-linux-gnu/libdl.so.2 (0xb7198000)
        libffi.so.6 => /usr/lib/i386-linux-gnu/libffi.so.6 (0xb7191000)
        libpcre.so.3 => /lib/i386-linux-gnu/libpcre.so.3 (0xb7155000)
        /lib/ld-linux.so.2 (0xb7730000)
        libz.so.1 => /lib/i386-linux-gnu/libz.so.1 (0xb713e000)
milan@milan:~$
```

Figure 12-3. *Using the ldd utility*

Before using ldd it is important to be aware of its limitations:

- ldd cannot identify the libraries dynamically loaded at runtime by calling the dlopen() function. In order to obtain this kind of information, different approaches must be applied. For more details, please visit Chapter 13.

- According to its man page, running certain ldd versions may actually represent a security threat.

Safer ldd Alternatives

As stated in the man page:

> *Be aware, however, that in some circumstances, some versions of ldd may attempt to obtain the dependency information by directly executing the program. Thus, you should never employ ldd on an untrusted executable, since this may result in the execution of arbitrary code. A safer alternative when dealing with untrusted executables is* the following (and also shown in Figure 12-4):

```
$ objdump -p /path/to/program | grep NEEDED
```

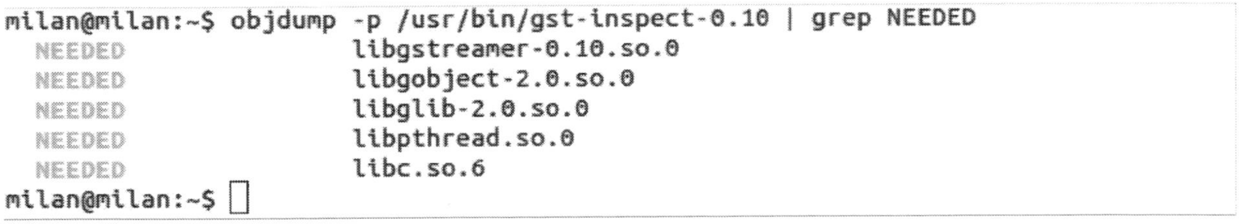

```
milan@milan:~$ objdump -p /usr/bin/gst-inspect-0.10 | grep NEEDED
  NEEDED                   libgstreamer-0.10.so.0
  NEEDED                   libgobject-2.0.so.0
  NEEDED                   libglib-2.0.so.0
  NEEDED                   libpthread.so.0
  NEEDED                   libc.so.6
milan@milan:~$ 
```

Figure 12-4. *Using objdump to (only partially) substitute the ldd utility*

The same result may be achieved by using the readelf utility (Figure 12-5):

```
$ readelf -d /path/to/program | grep NEEDED
```

```
milan@milan:~$ readelf -d /usr/bin/gst-inspect-0.10 | grep NEEDED
 0x00000001 (NEEDED)                     Shared library: [libgstreamer-0.10.so.0]
 0x00000001 (NEEDED)                     Shared library: [libgobject-2.0.so.0]
 0x00000001 (NEEDED)                     Shared library: [libglib-2.0.so.0]
 0x00000001 (NEEDED)                     Shared library: [libpthread.so.0]
 0x00000001 (NEEDED)                     Shared library: [libc.so.6]
milan@milan:~$
```

Figure 12-5. Using readelf to (only partially) substitute the ldd utility

Obviously, in the analysis of dependencies both tools do not go deeper than merely reading out the list of the most immediate dependencies from the binary file. From a security standpoint, this is definitely a safer method of finding the answer.

However, the provided list is nowhere near being as exhaustively complete as typically provided by ldd. In order to match it, you would probably need to conduct the recursive search on your own.

nm

The nm utility (http://linux.die.net/man/1/nm) is used to list the symbols of a binary file (Figure 12-6). The output line that prints out the symbol also indicates the symbol type. If the binary contains C++ code, the symbols are printed by default in the mangled form. Here are some of the most typically used input argument combinations:

- $ nm <path-to-binary> lists all symbols of a binary file. In case of shared libraries, it means not only the exported (of the .dynamic section) but all other symbols as well. If the library has been stripped (by using the strip command), nm without arguments will report no symbols found.

- $ nm -D <path-to-binary> lists only the symbols in the dynamic section (i.e., exported/ visible symbols of a shared library).

- $ nm -C <path-to-binary lists the symbols in demangled format (Figure 12-6).

```
00084ac0 T pspell_aspell_dummy()
00094040 T acommon::BetterList::set_cur_rank()
00093ed0 T acommon::BetterList::set_best_from_cur()
00094000 T acommon::BetterList::init()
000945f0 T acommon::BetterList::BetterList()
000945f0 T acommon::BetterList::BetterList()
00096c20 W acommon::BetterList::~BetterList()
00096af0 W acommon::BetterList::~BetterList()
00096af0 W acommon::BetterList::~BetterList()
00093f30 T acommon::BetterSize::set_cur_rank()
00093f10 T acommon::BetterSize::set_best_from_cur()
00093ef0 T acommon::BetterSize::init()
00096aa0 W acommon::BetterSize::~BetterSize()
00096a70 W acommon::BetterSize::~BetterSize()
00096a70 W acommon::BetterSize::~BetterSize()
0002dd50 W acommon::BlockSList<acommon::StringPair>::clear()
0002df50 W acommon::BlockSList<acommon::StringPair>::add_block(unsigned int)
0005da00 W acommon::BlockSList<acommon::String>::add_block(unsigned int)
00082900 W acommon::BlockSList<aspeller::Conds const*>::add_block(unsigned int)
```

Figure 12-6. Using the nm utility to list unmangled symbols

- `$ nm -D --no-demangle <path-to-binary>` prints the dynamic symbols of shared library and strictly requires that the symbols are not demangled (Figure 12-7).

```
00084ac0 T _Z19pspell_aspell_dummyv
00094040 T _ZN7acommon10BetterList12set_cur_rankEv
00093ed0 T _ZN7acommon10BetterList17set_best_from_curEv
00094000 T _ZN7acommon10BetterList4initEv
000945f0 T _ZN7acommon10BetterListC1Ev
000945f0 T _ZN7acommon10BetterListC2Ev
00096c20 W _ZN7acommon10BetterListD0Ev
00096af0 W _ZN7acommon10BetterListD1Ev
00096af0 W _ZN7acommon10BetterListD2Ev
00093f30 T _ZN7acommon10BetterSize12set_cur_rankEv
00093f10 T _ZN7acommon10BetterSize17set_best_from_curEv
00093ef0 T _ZN7acommon10BetterSize4initEv
00096aa0 W _ZN7acommon10BetterSizeD0Ev
00096a70 W _ZN7acommon10BetterSizeD1Ev
00096a70 W _ZN7acommon10BetterSizeD2Ev
0002dd50 W _ZN7acommon10BlockSListINS_10StringPairEE5clearEv
0002df50 W _ZN7acommon10BlockSListINS_10StringPairEE9add_blockEj
0005da00 W _ZN7acommon10BlockSListINS_6StringEE9add_blockEj
00082900 W _ZN7acommon10BlockSListIPKN8aspeller5CondsEE9add_blockEj
0005d740 W _ZN7acommon10BlockSListIPKcE9add_blockEj
```

Figure 12-7. Using the nm utility to list mangled symbols

This option is extremely useful in detecting the most usual bug when designing the shared library—the case when the designer forgets the extern "C" specifier in the ABI function declaration/definition (which happens to be exactly what the client binary expects to find).

- `$ nm -A <library-folder-path>/* | grep symbol-name` is useful when you search for a symbol in multitude of binaries located in the same folder, as -A option prints the name of each library in which a symbols is found (Figure 12-8).

```
milan@milan-ub-1204-32-lts:/usr/lib$ nm -DA * | grep pspell_aspell_dummy
                    ○
                    ○
                    ○
libaspell.so.15:00084ac0 T _Z19pspell_aspell_dummyv
libaspell.so.15.2.0:00084ac0 T _Z19pspell_aspell_dummyv
libpspell.so.15:00000430 T _Z19pspell_aspell_dummyv
libpspell.so.15.2.0:00000430 T _Z19pspell_aspell_dummyv
```

Figure 12-8. Using nm to recursively search for the presence of a symbol in the set of libraries

- `$ nm -u <path-to-binary>` is useful when you want to list the library's undefined symbols (i.e., the symbols that the library itself does not contain, but counts on to be provided at runtime, possibly by some other loaded dynamic library).

The web article at www.thegeekstuff.com/2012/03/linux-nm-command/ lists the 10 most useful nm commands.

objdump

The objdump (http://linux.die.net/man/1/objdump) utility program is probably the single most versatile binary analysis tool. Chronologically, it is older than readelf, which parallels its abilities in many cases. The advantage of objdump is that in addition to ELF, it supports about 50 other binary formats. Also, its disassembling capabilities are better than those of readelf.

The following sections cover the tasks that most frequently make use of objdump.

Parsing ELF Header

The objdump **-f** command-line option is used to obtain an insight into the object file's header. The header provides plenty of useful information. In particular, the binary type (object file/static library vs. dynamic library vs. executable) as well as the information about the entry point (start of the .text section) may be quickly obtained (Figure 12-9).

```
milan@milan$ objdump -f ./driverApp/driver

./driverApp/driver:      file format elf32-i386
architecture: i386, flags 0x00000112:
EXEC_P, HAS_SYMS, D_PAGED
start address 0x080484c0
```
```
milan@milan$ objdump -f ./sharedLib/libmreloc.so

./sharedLib/libmreloc.so:      file format elf32-i386
architecture: i386, flags 0x00000150:
HAS_SYMS, DYNAMIC, D_PAGED
start address 0x00000390
```
```
milan@milan$ objdump -f ./ml_mainreloc.o

./ml_mainreloc.o:      file format elf32-i386
architecture: i386, flags 0x00000011:
HAS_RELOC, HAS_SYMS
start address 0x00000000
```

Figure 12-9. *Using objdump to parse the ELF header of various binary file types*

When examining the static library, objdump -f prints out the header of each and every object file found in the library.

Listing and Examining Sections

The objdump -h option is used to list the available sections (Figure 12-10).

```
milan@milan$ objdump -h libmreloc.so

libmreloc.so:     file format elf32-i386

Sections:
Idx Name          Size      VMA       LMA       File off  Algn
  0 .note.gnu.build-id 00000024  00000114  00000114  00000114  2**2
                  CONTENTS, ALLOC, LOAD, READONLY, DATA
  1 .gnu.hash     00000040  00000138  00000138  00000138  2**2
                  CONTENTS, ALLOC, LOAD, READONLY, DATA
  2 .dynsym       000000b0  00000178  00000178  00000178  2**2
                  CONTENTS, ALLOC, LOAD, READONLY, DATA
  3 .dynstr       0000007c  00000220  00000228  00000228  2**0
                  CONTENTS, ALLOC, LOAD, READONLY, DATA
  4 .gnu.version  00000016  000002a4  000002a4  000002a4  2**1
                  CONTENTS, ALLOC, LOAD, READONLY, DATA
  5 .gnu.version_r 00000020  000002bc  000002bc  000002bc  2**2
                  CONTENTS, ALLOC, LOAD, READONLY, DATA
  6 .rel.dyn      00000038  000002dc  000002dc  000002dc  2**2
                  CONTENTS, ALLOC, LOAD, READONLY, DATA
  7 .rel.plt      00000010  00000314  00000314  00000314  2**2
                  CONTENTS, ALLOC, LOAD, READONLY, DATA
  8 .init         0000002e  00000324  00000324  00000324  2**2
                  CONTENTS, ALLOC, LOAD, READONLY, CODE
  9 .plt          00000030  00000360  00000360  00000360  2**4
                  CONTENTS, ALLOC, LOAD, READONLY, CODE
 10 .text         00000118  00000390  00000390  00000390  2**4
                  CONTENTS, ALLOC, LOAD, READONLY, CODE
 11 .fini         0000001a  000004a8  000004a8  000004a8  2**2
                                ○
                                ○
                                ○
                                ○
 21 .bss          00000008  00002010  00002010  00001010  2**2
                  ALLOC
 22 .comment      0000002a  00000000  00000000  00001010  2**0
                  CONTENTS, READONLY
 23 .debug_aranges 00000020  00000000  00000000  0000103a  2**0
                  CONTENTS, READONLY, DEBUGGING
 24 .debug_info   00000075  00000000  00000000  0000105a  2**0
                  CONTENTS, READONLY, DEBUGGING
 25 .debug_abbrev 00000058  00000000  00000000  000010cf  2**0
                  CONTENTS, READONLY, DEBUGGING
 26 .debug_line   0000003f  00000000  00000000  00001127  2**0
                  CONTENTS, READONLY, DEBUGGING
 27 .debug_str    0000004c  00000000  00000000  00001166  2**0
                  CONTENTS, READONLY, DEBUGGING
 28 .debug_loc    00000038  00000000  00000000  000011b2  2**0
                  CONTENTS, READONLY, DEBUGGING
milan@milan$
```

Figure 12-10. *Using objdump to list the binary file sections*

When it comes to section examinations, objdump provides dedicated command switches for the sections that are most frequently topic of interest for the programmers. In the following sections, I look at some of the notable examples.

Listing All Symbols

Running objdump -t <path-to-binary> provides output that is the full equivalent of running nm <path-to-binary> (Figure 12-11).

```
milan@milan$ objdump -t libdemo1.so

libdemo1.so:     file format elf32-i386

SYMBOL TABLE:
00000114 l    d  .note.gnu.build-id    00000000              .note.gnu.build-id
00000138 l    d  .gnu.hash     00000000           .gnu.hash
00000174 l    d  .dynsym       00000000           .dynsym
00000224 l    d  .dynstr       00000000           .dynstr
000002b6 l    d  .gnu.version  00000000           .gnu.version
000002cc l    d  .gnu.version_r 00000000          .gnu.version_r
000002fc l    d  .rel.dyn      00000000           .rel.dyn
0000032c l    d  .rel.plt      00000000           .rel.plt
0000033c l    d  .init 00000000          .init
00000370 l    d  .plt  00000000          .plt
000003a0 l    d  .text 00000000          .text
000004b8 l    d  .fini 00000000          .fini
000004d4 l    d  .rodata       00000000           .rodata
000004f8 l    d  .eh_frame_hdr 00000000           .eh_frame_hdr
00000514 l    d  .eh_frame     00000000           .eh_frame
00001f0c l    d  .ctors 00000000         .ctors
00001f14 l    d  .dtors 00000000         .dtors
00001f1c l    d  .jcr  00000000          .jcr
00001f20 l    d  .dynamic      00000000           .dynamic
00001fe8 l    d  .got  00000000          .got
00001ff4 l    d  .got.plt      00000000           .got.plt
00002008 l    d  .data 00000000          .data
0000200c l    d  .bss  00000000          .bss
00000000 l    d  .comment      00000000           .comment
00000000 l    d  .debug_aranges 00000000          .debug_aranges
00000000 l    d  .debug_info   00000000           .debug_info
00000000 l    d  .debug_abbrev 00000000           .debug_abbrev
00000000 l    d  .debug_line   00000000           .debug_line
00000000 l    d  .debug_str    00000000           .debug_str
00000000 l    d  .debug_loc    00000000           .debug_loc
00000000 l    df *ABS*  00000000         crtstuff.c
00001f0c l    O .ctors 00000000          __CTOR_LIST__
00001f14 l    O .dtors 00000000          __DTOR_LIST__
00001f1c l    O .jcr  00000000           __JCR_LIST__
000003a0 l    F .text 00000000          __do_global_dtors_aux
0000200c l    O .bss  00000001           completed.6159
00002010 l    O .bss  00000004           dtor_idx.6161
00000420 l    F .text 00000000          frame_dummy
00000000 l    df *ABS*  00000000         crtstuff.c
00001f10 l    O .ctors 00000000          __CTOR_END__
00000570 l    O .eh_frame     00000000              __FRAME_END__
00001f1c l    O .jcr  00000000           __JCR_END__
00000480 l    F .text 00000000          __do_global_ctors_aux
00000000 l    df *ABS*  00000000         sharedLib1Functions.c
00000457 l    F .text 00000000          __i686.get_pc_thunk.bx
00001f18 l    O .dtors 00000000          __DTOR_END__
00002008 l    O .data 00000000          __dso_handle
00001f20 l    O *ABS*  00000000         _DYNAMIC
00001ff4 l    O *ABS*  00000000         _GLOBAL_OFFSET_TABLE_
00000000      F *UND*  00000000         printf@@GLIBC_2.0
0000200c g       *ABS*  00000000         _edata
000004b8 g    F .fini 00000000          _fini
00000000  w   F *UND*  00000000         __cxa_finalize@@GLIBC_2.1.3
00000000  w     *UND*  00000000         __gmon_start__
00002014 g       *ABS*  00000000         _end
0000200c g       *ABS*  00000000         __bss_start
00000000  w     *UND*  00000000         _Jv_RegisterClasses
0000045c g    F .text 0000001c         sharedLib1Function
0000033c g    F .init 00000000          _init
milan@milan$
```

Figure 12-11. *Using objdump to list all symbols*

Listing Dynamic Symbols Only

Running objdump -T <path-to-binary> provides output that is the full equivalent of running nm -D <path-to-binary> (Figure 12-12).

```
milan@milan$ objdump -T libdemo1.so

libdemo1.so:     file format elf32-i386

DYNAMIC SYMBOL TABLE:
00000000      DF *UND*  00000000  GLIBC_2.0   printf
00000000  w   DF *UND*  00000000  GLIBC_2.1.3 __cxa_finalize
00000000  w   D  *UND*  00000000              __gmon_start__
00000000  w   D  *UND*  00000000              _Jv_RegisterClasses
0000200c  g   D  *ABS*  00000000  Base        _edata
00002014  g   D  *ABS*  00000000  Base        _end
0000200c  g   D  *ABS*  00000000  Base        __bss_start
0000033c  g   DF .init  00000000  Base        _init
000004b8  g   DF .fini  00000000  Base        _fini
0000045c  g   DF .text  0000001c  Base        sharedLib1Function
```

Figure 12-12. Using objdump to list only dynamic symbols

Examining Dynamic Section

Running objdump -p <path-to-binary> examines the dynamic section (useful for finding DT_RPATH and/or DT_RUNPATH settings). Please notice that in this scenario you care about the final part of displayed output (Figure 12-13).

```
milan@milan$ objdump -p demoApp

demoApp:      file format elf64-x86-64

Program Header:
    PHDR off     0x0000000000000040 vaddr 0x00000000003ff040 paddr 0x00000000003ff040 alig
                          ○
                          ○
                          ○
                          ○
Dynamic Section:
  NEEDED                 libpthread.so.0
  NEEDED                 libdl.so.2
  NEEDED                 libdynamiclinkingdemo.so
  NEEDED                 libstdc++.so.6
  NEEDED                 libm.so.6
  NEEDED                 libgcc_s.so.1
  NEEDED                 libc.so.6
  RUNPATH                ../deploy:./deploy
  INIT                   0x00000000004005d8
  FINI                   0x0000000000400808
  HASH                   0x00000000003ff4d0
  GNU_HASH               0x00000000003ff490
  STRTAB                 0x00000000003ff270
  SYMTAB                 0x00000000003ff388
  STRSZ                  0x0000000000000115
  SYMENT                 0x0000000000000018
  DEBUG                  0x0000000000000000
  PLTGOT                 0x0000000000600fe8
  PLTRELSZ               0x0000000000000048
  PLTREL                 0x0000000000000007
  JMPREL                 0x0000000000400590
  RELA                   0x0000000000400578
  RELASZ                 0x0000000000000018
  RELAENT                0x0000000000000018
  VERNEED                0x0000000000400538
  VERNEEDNUM             0x0000000000000002
  VERSYM                 0x0000000000400522

Version References:
  required from libstdc++.so.6:
    0x056bafd3 0x00 03 CXXABI_1.3
  required from libc.so.6:
    0x09691a75 0x00 02 GLIBC_2.2.5

milan@milan$
```

Figure 12-13. *Using objdump to examine the library dynamic section*

Examining Relocation Section

Running objdump -R <path-to-binary> examines the relocation section (Figure 12-14).

```
milan@milan$ objdump -R sharedLib/libmreloc.so

sharedLib/libmreloc.so:     file format elf32-i386

DYNAMIC RELOCATION RECORDS
OFFSET    TYPE                 VALUE
00002008 R_386_RELATIVE      *ABS*
00000450 R_386_32            myglob
00000458 R_386_32            myglob
0000045d R_386_32            myglob
00001fe8 R_386_GLOB_DAT       __cxa_finalize
00001fec R_386_GLOB_DAT       __gmon_start__
00001ff0 R_386_GLOB_DAT       _Jv_RegisterClasses
00002000 R_386_JUMP_SLOT      __cxa_finalize
00002004 R_386_JUMP_SLOT      __gmon_start__
```

Figure 12-14. *Using objdump to list the relocation section*

Examining Data Section

Running objdump -s -j <section name> <path-to-binary> provides the hex dump of the values carried by the section. In Figure 12-15, it is the .got section.

```
milan@milan$ objdump -s -j .got driver

driver:     file format elf32-i386

Contents of section .got:
 8049ff0 00000000
milan@milan$
```

Figure 12-15. *Using objdump to examine the data section*

Listing and Examining Segments

Running objdump -p <path-to-binary> displays information about the ELF binary segments. Note that only the first part of the displayed output pertains to this particular task (Figure 12-16).

```
milan@milan$ objdump -p demoApp

demoApp:     file format elf64-x86-64

Program Header:
    PHDR off    0x0000000000000040 vaddr 0x00000000003ff040 paddr 0x00000000003ff040 align 2**3
         filesz 0x0000000000000230 memsz 0x0000000000000230 flags r-x
   STACK off    0x0000000000001000 vaddr 0x0000000000000000 paddr 0x0000000000000000 align 2**3
         filesz 0x0000000000000000 memsz 0x0000000000000000 flags rw-
    LOAD off    0x0000000000000000 vaddr 0x00000000003ff000 paddr 0x00000000003ff000 align 2**12
         filesz 0x0000000000001000 memsz 0x0000000000001000 flags rw-
  INTERP off    0x0000000000000510 vaddr 0x00000000003ff510 paddr 0x00000000003ff510 align 2**0
         filesz 0x000000000000001c memsz 0x000000000000001c flags r--
    LOAD off    0x0000000000001000 vaddr 0x0000000000400000 paddr 0x0000000000400000 align 2**12
         filesz 0x00000000000008dc memsz 0x00000000000008dc flags r-x
    NOTE off    0x0000000000001254 vaddr 0x0000000000400254 paddr 0x0000000000400254 align 2**2
         filesz 0x0000000000000044 memsz 0x0000000000000044 flags r--
EH_FRAME off    0x000000000000181c vaddr 0x000000000040081c paddr 0x000000000040081c align 2**2
         filesz 0x0000000000000024 memsz 0x0000000000000024 flags r--
    LOAD off    0x0000000000001da8 vaddr 0x0000000000600da8 paddr 0x0000000000600da8 align 2**12
         filesz 0x0000000000000280 memsz 0x0000000000000290 flags rw-
   RELRO off    0x0000000000001da8 vaddr 0x0000000000600da8 paddr 0x0000000000600da8 align 2**0
         filesz 0x0000000000000258 memsz 0x0000000000000258 flags r--
 DYNAMIC off    0x0000000000001dd0 vaddr 0x0000000000600dd0 paddr 0x0000000000600dd0 align 2**3
         filesz 0x0000000000000210 memsz 0x0000000000000210 flags rw-
                                    o
                                    o
                                    o
                                    o
```

Figure 12-16. *Using objdump to list segments*

Disassembling the Code

Here are a few examples of how objdump may be used to disassemble the code:

- Disassembling and specifying assembler notation flavor (Intel style in this case), as shown in Figure 12-17.

```
milan@milan$ objdump -d -Mintel libmreloc.so | grep -A 10 ml_
0000044c <ml_func>:
 44c:   55                      push    ebp
 44d:   89 e5                   mov     ebp,esp
 44f:   a1 00 00 00 00          mov     eax,ds:0x0
 454:   03 45 08                add     eax,DWORD PTR [ebp+0x8]
 457:   a3 00 00 00 00          mov     ds:0x0,eax
 45c:   a1 00 00 00 00          mov     eax,ds:0x0
 461:   03 45 0c                add     eax,DWORD PTR [ebp+0xc]
 464:   5d                      pop     ebp
 465:   c3                      ret
 466:   90                      nop
milan@milan$
milan@milan$ objdump -d -Mintel driver | grep -A 10 "<main>"
08048646 <main>:
 8048646:       55                      push    ebp
 8048647:       89 e5                   mov     ebp,esp
 8048649:       83 e4 f0                and     esp,0xfffffff0
 804864c:       83 ec 20                sub     esp,0x20
 804864f:       c7 44 24 04 00 00 00    mov     DWORD PTR [esp+0x4],0x0
 8048656:       00
 8048657:       c7 04 24 74 85 04 08    mov     DWORD PTR [esp],0x8048574
 804865e:       e8 2d fe ff ff          call    8048490 <dl_iterate_phdr@plt>
 8048663:       8b 45 08                mov     eax,DWORD PTR [ebp+0x8]
 8048666:       89 44 24 04             mov     DWORD PTR [esp+0x4],eax
milan@milan$
```

Figure 12-17. *Using objdump to disassemble the binary file*

- Disassembling and Intel style and interspersing the original source code (Figure 12-18).

```
milan@milan$ objdump -d -S -M intel ./libdemo1.so | grep -A 26 "<sharedLib1Function>"
0000045c <sharedLib1Function>:
#include "sharedLib1Functions.h"

void sharedLib1Function(int x)
{
 45c:   55                      push    ebp
 45d:   89 e5                   mov     ebp,esp
 45f:   83 ec 18                sub     esp,0x18
        printf("sharedLib1Function(%d) is called\n", x);
 462:   b8 d4 04 00 00          mov     eax,0x4d4
 467:   8b 55 08                mov     edx,DWORD PTR [ebp+0x8]
 46a:   89 54 24 04             mov     DWORD PTR [esp+0x4],edx
 46e:   89 04 24                mov     DWORD PTR [esp],eax
 471:   e8 fc ff ff ff          call    472 <sharedLib1Function+0x16>
}
 476:   c9                      leave
 477:   c3                      ret
 478:   90                      nop
 479:   90                      nop
 47a:   90                      nop
 47b:   90                      nop
 47c:   90                      nop
 47d:   90                      nop
 47e:   90                      nop
 47f:   90                      nop

00000480 <__do_global_ctors_aux>:
milan@milan$
```

Figure 12-18. *Using objdump to disassemble the binary file (Intel syntax)*

This option works only if the binary is built for debug (i.e., with the -g option).

- Disassembling specific sections.

  Other than .text section that carries the code, the binary may contain other sections (.plt, for example) that also contain the code. By default, objdump disassembles all the sections carrying the code. However, there may be scenarios in which you are interested in examining the code carried strictly by a given section (Figure 12-19).

```
milan@milan$ objdump -d -M intel -j .plt driver

driver:     file format elf32-i386


Disassembly of section .plt:

08048470 <strstr@plt-0x10>:
 8048470:       ff 35 f8 9f 04 08       push    DWORD PTR ds:0x8049ff8
 8048476:       ff 25 fc 9f 04 08       jmp     DWORD PTR ds:0x8049ffc
 804847c:       00 00                   add     BYTE PTR [eax],al
        ...

08048480 <strstr@plt>:
 8048480:       ff 25 00 a0 04 08       jmp     DWORD PTR ds:0x804a000
 8048486:       68 00 00 00 00          push    0x0
 804848b:       e9 e0 ff ff ff          jmp     8048470 <_init+0x38>

08048490 <printf@plt>:
 8048490:       ff 25 04 a0 04 08       jmp     DWORD PTR ds:0x804a004
 8048496:       68 08 00 00 00          push    0x8
 804849b:       e9 d0 ff ff ff          jmp     8048470 <_init+0x38>

080484a0 <ml_func@plt>:
 80484a0:       ff 25 08 a0 04 08       jmp     DWORD PTR ds:0x804a008
 80484a6:       68 10 00 00 00          push    0x10
 80484ab:       e9 c0 ff ff ff          jmp     8048470 <_init+0x38>

080484b0 <__gmon_start__@plt>:
 80484b0:       ff 25 0c a0 04 08       jmp     DWORD PTR ds:0x804a00c
 80484b6:       68 18 00 00 00          push    0x18
 80484bb:       e9 b0 ff ff ff          jmp     8048470 <_init+0x38>

080484c0 <dl_iterate_phdr@plt>:
 80484c0:       ff 25 10 a0 04 08       jmp     DWORD PTR ds:0x804a010
 80484c6:       68 20 00 00 00          push    0x20
 80484cb:       e9 a0 ff ff ff          jmp     8048470 <_init+0x38>

080484d0 <__libc_start_main@plt>:
 80484d0:       ff 25 14 a0 04 08       jmp     DWORD PTR ds:0x804a014
 80484d6:       68 28 00 00 00          push    0x28
 80484db:       e9 90 ff ff ff          jmp     8048470 <_init+0x38>

080484e0 <putchar@plt>:
 80484e0:       ff 25 18 a0 04 08       jmp     DWORD PTR ds:0x804a018
 80484e6:       68 30 00 00 00          push    0x30
 80484eb:       e9 80 ff ff ff          jmp     8048470 <_init+0x38>
milan@milan$
```

Figure 12-19. Using objdump to disassemble a specific section

objdump nm equivalents

objdump can be used to provide full equivalents of the nm command:

- $ nm <path-to-binary>

  equivalent is

  $ objdump -t <path-to-binary>

- $ nm -D <path-to-binary>

  equivalent is

  $ objdump -T <path-to-binary>

- $ nm -C <path-to-binary>

  equivalent is

  $ objdump -C <path-to-binary>

readelf

The readelf (http://linux.die.net/man/1/readelf) command-line utility provides almost completely duplicate functionality found in the objdump utility. The most notable differences between the readelf and objdump are

- readelf supports only ELF binary format. On the other hand, the objdump can analyze about 50 different binary formats, including the Windows PE/COFF format.

- readelf does not depend on the Binary File Descriptor library (http://en.wikipedia.org/wiki/Binary_File_Descriptor_library), which all GNU object file parsing tools depend on, thus providing the independent insight into the contents of ELF format

The following two sections provide an overview of the most common tasks that make use of objdump.

Parsing ELF Header

The readelf -h command-line option is used to obtain an insight into the object file's header. The header provides plenty of useful information. In particular, the binary type (object file/static library vs. dynamic library vs. executable) as well as the information about the entry point (the start of the .text section) may be quickly obtained (Figure 12-20).

```
milan@milan$ readelf -h driverApp/driver
ELF Header:
  Magic:   7f 45 4c 46 01 01 01 00 00 00 00 00 00 00 00 00
  Class:                             ELF32
  Data:                              2's complement, little endian
  Version:                           1 (current)
  OS/ABI:                            UNIX - System V
  ABI Version:                       0
  Type:                              EXEC (Executable file)
  Machine:                           Intel 80386
  Version:                           0x1
  Entry point address:               0x80484c0
  Start of program headers:          52 (bytes into file)
  Start of section headers:          6196 (bytes into file)
  Flags:                             0x0
  Size of this header:               52 (bytes)
  Size of program headers:           32 (bytes)
  Number of program headers:         9
  Size of section headers:           40 (bytes)
  Number of section headers:         36
  Section header string table index: 33
milan@milan$
```

```
milan@milan$ readelf -h sharedLib/libmreloc.so
ELF Header:
  Magic:   7f 45 4c 46 01 01 01 00 00 00 00 00 00 00 00 00
  Class:                             ELF32
  Data:                              2's complement, little endian
  Version:                           1 (current)
  OS/ABI:                            UNIX - System V
  ABI Version:                       0
  Type:                              DYN (Shared object file)
  Machine:                           Intel 80386
  Version:                           0x1
  Entry point address:               0x390
  Start of program headers:          52 (bytes into file)
  Start of section headers:          4884 (bytes into file)
  Flags:                             0x0
  Size of this header:               52 (bytes)
  Size of program headers:           32 (bytes)
  Number of program headers:         7
  Size of section headers:           40 (bytes)
  Number of section headers:         33
  Section header string table index: 30
milan@milan$
```

Figure 12-20. (*continued*)

```
milan@milan$ readelf -h ml_mainreloc.o
ELF Header:
  Magic:   7f 45 4c 46 01 01 01 00 00 00 00 00 00 00 00 00
  Class:                             ELF32
  Data:                              2's complement, little endian
  Version:                           1 (current)
  OS/ABI:                            UNIX - System V
  ABI Version:                       0
  Type:                              REL (Relocatable file)
  Machine:                           Intel 80386
  Version:                           0x1
  Entry point address:               0x0
  Start of program headers:          0 (bytes into file)
  Start of section headers:          832 (bytes into file)
  Flags:                             0x0
  Size of this header:               52 (bytes)
  Size of program headers:           0 (bytes)
  Number of program headers:         0
  Size of section headers:           40 (bytes)
  Number of section headers:         21
  Section header string table index: 18
milan@milan$
```

Figure 12-20. *Examples of using readelf to examine the ELF header of executable, shared library, and object file/static library*

When examining the static library, readelf -h prints out the header of each and every object file found in the library.

Listing and Examining Sections

The readelf -S option is used to list the available sections (Figure 12-21).

```
milan@milan$ readelf -S libmreloc.so
There are 33 section headers, starting at offset 0x1314:

Section Headers:
  [Nr] Name              Type      Addr     Off    Size   ES Flg Lk Inf Al
  [ 0]                   NULL      00000000 000000 000000 00      0   0  0
  [ 1] .note.gnu.build-i NOTE      00000114 000114 000024 00   A  0   0  4
  [ 2] .gnu.hash         GNU_HASH  00000138 000138 000040 04   A  3   0  4
  [ 3] .dynsym           DYNSYM    00000178 000178 0000b0 10   A  4   1  4
  [ 4] .dynstr           STRTAB    00000228 000228 00007c 00   A  0   0  1
  [ 5] .gnu.version      VERSYM    000002a4 0002a4 000016 02   A  3   0  2
  [ 6] .gnu.version_r    VERNEED   000002bc 0002bc 000020 00   A  4   1  4
  [ 7] .rel.dyn          REL       000002dc 0002dc 000038 08   A  3   0  4
  [ 8] .rel.plt          REL       00000314 000314 000010 08   A  3  10  4
  [ 9] .init             PROGBITS  00000324 000324 00002e 00  AX  0   0  4
  [10] .plt              PROGBITS  00000360 000360 000030 04  AX  0   0 16
  [11] .text             PROGBITS  00000390 000390 000118 00  AX  0   0 16
  [12] .fini             PROGBITS  000004a8 0004a8 00001a 00  AX  0   0  4
  [13] .eh_frame_hdr     PROGBITS  000004c4 0004c4 00001c 00   A  0   0  4
  [14] .eh_frame         PROGBITS  000004e0 0004e0 000060 00   A  0   0  4
  [15] .ctors            PROGBITS  00001f0c 000f0c 000008 00  WA  0   0  4
  [16] .dtors            PROGBITS  00001f14 000f14 000008 00  WA  0   0  4
  [17] .jcr              PROGBITS  00001f1c 000f1c 000004 00  WA  0   0  4
  [18] .dynamic          DYNAMIC   00001f20 000f20 0000c8 08  WA  4   0  4
  [19] .got              PROGBITS  00001fe8 000fe8 00000c 04  WA  0   0  4
  [20] .got.plt          PROGBITS  00001ff4 000ff4 000014 04  WA  0   0  4
  [21] .data             PROGBITS  00002008 001008 000008 00  WA  0   0  4
  [22] .bss              NOBITS    00002010 001010 000008 00  WA  0   0  4
  [23] .comment          PROGBITS  00000000 001010 00002a 01  MS  0   0  1
  [24] .debug_aranges    PROGBITS  00000000 00103a 000020 00      0   0  1
  [25] .debug_info       PROGBITS  00000000 00105a 000075 00      0   0  1
  [26] .debug_abbrev     PROGBITS  00000000 0010cf 000058 00      0   0  1
  [27] .debug_line       PROGBITS  00000000 001127 00003f 00      0   0  1
  [28] .debug_str        PROGBITS  00000000 001166 00004c 01  MS  0   0  1
  [29] .debug_loc        PROGBITS  00000000 0011b2 000038 00      0   0  1
  [30] .shstrtab         STRTAB    00000000 0011ea 000129 00      0   0  1
  [31] .symtab           SYMTAB    00000000 00183c 0003b0 10     32  49  4
  [32] .strtab           STRTAB    00000000 001bec 000182 00      0   0  1
Key to Flags:
  W (write), A (alloc), X (execute), M (merge), S (strings)
  I (info), L (link order), G (group), T (TLS), E (exclude), x (unknown)
  O (extra OS processing required) o (OS specific), p (processor specific)
```

Figure 12-21. Using readelf to list sections

When it comes to section examinations, readelf provides dedicated command switches for the sections that are most frequently topics of interest for programmers, such as the .symtab, .dynsym, and .dynamic sections.

Listing All Symbols

Running readelf --symbols provides output that is the full equivalent of running nm <path-to-binary> (Figure 12-22).

```
milan@milan$ readelf --symbols libdemo1.so

Symbol table '.dynsym' contains 11 entries:
   Num:    Value  Size Type    Bind   Vis      Ndx Name
     0: 00000000     0 NOTYPE  LOCAL  DEFAULT  UND
     1: 00000000     0 FUNC    GLOBAL DEFAULT  UND printf@GLIBC_2.0 (2)
     2: 00000000     0 FUNC    WEAK   DEFAULT  UND __cxa_finalize@GLIBC_2.1.3 (3)
     3: 00000000     0 NOTYPE  WEAK   DEFAULT  UND __gmon_start__
     4: 00000000     0 NOTYPE  WEAK   DEFAULT  UND _Jv_RegisterClasses
     5: 0000200c     0 NOTYPE  GLOBAL DEFAULT  ABS _edata
     6: 00002014     0 NOTYPE  GLOBAL DEFAULT  ABS _end
     7: 0000200c     0 NOTYPE  GLOBAL DEFAULT  ABS __bss_start
     8: 0000033c     0 FUNC    GLOBAL DEFAULT    9 _init
     9: 000004b8     0 FUNC    GLOBAL DEFAULT   12 _fini
    10: 0000045c    28 FUNC    GLOBAL DEFAULT   11 sharedLib1Function

Symbol table '.symtab' contains 60 entries:
   Num:    Value  Size Type    Bind   Vis      Ndx Name
     0: 00000000     0 NOTYPE  LOCAL  DEFAULT  UND
     1: 00000114     0 SECTION LOCAL  DEFAULT    1
     2: 00000138     0 SECTION LOCAL  DEFAULT    2
     3: 00000174     0 SECTION LOCAL  DEFAULT    3
     4: 00000224     0 SECTION LOCAL  DEFAULT    4
     5: 000002b6     0 SECTION LOCAL  DEFAULT    5
     6: 000002cc     0 SECTION LOCAL  DEFAULT    6
     7: 000002fc     0 SECTION LOCAL  DEFAULT    7
     8: 0000032c     0 SECTION LOCAL  DEFAULT    8
     9: 0000033c     0 SECTION LOCAL  DEFAULT    9
    10: 00000370     0 SECTION LOCAL  DEFAULT   10
    11: 000003a0     0 SECTION LOCAL  DEFAULT   11
    12: 000004b8     0 SECTION LOCAL  DEFAULT   12
    13: 000004d4     0 SECTION LOCAL  DEFAULT   13
    14: 000004f8     0 SECTION LOCAL  DEFAULT   14
    15: 00000514     0 SECTION LOCAL  DEFAULT   15
    16: 00001f0c     0 SECTION LOCAL  DEFAULT   16
    17: 00001f14     0 SECTION LOCAL  DEFAULT   17
    18: 00001f1c     0 SECTION LOCAL  DEFAULT   18
    19: 00001f20     0 SECTION LOCAL  DEFAULT   19
    20: 00001fe8     0 SECTION LOCAL  DEFAULT   20
    21: 00001ff4     0 SECTION LOCAL  DEFAULT   21
    22: 00002008     0 SECTION LOCAL  DEFAULT   22
    23: 0000200c     0 SECTION LOCAL  DEFAULT   23
    24: 00000000     0 SECTION LOCAL  DEFAULT   24
    25: 00000000     0 SECTION LOCAL  DEFAULT   25
    26: 00000000     0 SECTION LOCAL  DEFAULT   26
    27: 00000000     0 SECTION LOCAL  DEFAULT   27
    28: 00000000     0 SECTION LOCAL  DEFAULT   28
    29: 00000000     0 SECTION LOCAL  DEFAULT   29
    30: 00000000     0 SECTION LOCAL  DEFAULT   30
    31: 00000000     0 FILE    LOCAL  DEFAULT  ABS crtstuff.c
    32: 00001f0c     0 OBJECT  LOCAL  DEFAULT   16 __CTOR_LIST__
    33: 00001f14     0 OBJECT  LOCAL  DEFAULT   17 __DTOR_LIST__
    34: 00001f1c     0 OBJECT  LOCAL  DEFAULT   18 __JCR_LIST__
    35: 000003a0     0 FUNC    LOCAL  DEFAULT   11 __do_global_dtors_aux
    36: 0000200c     1 OBJECT  LOCAL  DEFAULT   23 completed.6159
    37: 00002010     4 OBJECT  LOCAL  DEFAULT   23 dtor_idx.6161
    38: 00000420     0 FUNC    LOCAL  DEFAULT   11 frame_dummy
    39: 00000000     0 FILE    LOCAL  DEFAULT  ABS crtstuff.c
    40: 00001f10     0 OBJECT  LOCAL  DEFAULT   16 __CTOR_END__
    41: 00000570     0 OBJECT  LOCAL  DEFAULT   15 __FRAME_END__
    42: 00001f1c     0 OBJECT  LOCAL  DEFAULT   18 __JCR_END__
    43: 00000480     0 FUNC    LOCAL  DEFAULT   11 __do_global_ctors_aux
    44: 00000000     0 FILE    LOCAL  DEFAULT  ABS sharedLib1Functions.c
    45: 00000457     0 FUNC    LOCAL  DEFAULT   11 __i686.get_pc_thunk.bx
    46: 00001f18     0 OBJECT  LOCAL  DEFAULT   17 __DTOR_END__
    47: 00002008     0 OBJECT  LOCAL  DEFAULT   22 __dso_handle
    48: 00001f20     0 OBJECT  LOCAL  DEFAULT  ABS _DYNAMIC
    49: 00001ff4     0 OBJECT  LOCAL  DEFAULT  ABS _GLOBAL_OFFSET_TABLE_
    50: 00000000     0 FUNC    GLOBAL DEFAULT  UND printf@@GLIBC_2.0
    51: 0000200c     0 NOTYPE  GLOBAL DEFAULT  ABS _edata
    52: 000004b8     0 FUNC    GLOBAL DEFAULT   12 _fini
    53: 00000000     0 FUNC    WEAK   DEFAULT  UND __cxa_finalize@@GLIBC_2.1
    54: 00000000     0 NOTYPE  WEAK   DEFAULT  UND __gmon_start__
    55: 00002014     0 NOTYPE  GLOBAL DEFAULT  ABS _end
    56: 0000200c     0 NOTYPE  GLOBAL DEFAULT  ABS __bss_start
    57: 00000000     0 NOTYPE  WEAK   DEFAULT  UND _Jv_RegisterClasses
    58: 0000045c    28 FUNC    GLOBAL DEFAULT   11 sharedLib1Function
    59: 0000033c     0 FUNC    GLOBAL DEFAULT    9 _init
milan@milan$
```

Figure 12-22. *Using readelf to list all symbols*

261

Listing Dynamic Symbols Only

Running readelf --dyn-syms provides output that is the full equivalent of running nm -D <path-to-binary> (Figure 12-23).

```
milan@milan$ readelf --dyn-syms libdemo1.so

Symbol table '.dynsym' contains 11 entries:
   Num:    Value  Size Type    Bind   Vis      Ndx Name
     0: 00000000     0 NOTYPE  LOCAL  DEFAULT  UND
     1: 00000000     0 FUNC    GLOBAL DEFAULT  UND printf@GLIBC_2.0 (2)
     2: 00000000     0 FUNC    WEAK   DEFAULT  UND __cxa_finalize@GLIBC_2.1.3 (3)
     3: 00000000     0 NOTYPE  WEAK   DEFAULT  UND __gmon_start__
     4: 00000000     0 NOTYPE  WEAK   DEFAULT  UND _Jv_RegisterClasses
     5: 0000200c     0 NOTYPE  GLOBAL DEFAULT  ABS _edata
     6: 00002014     0 NOTYPE  GLOBAL DEFAULT  ABS _end
     7: 0000200c     0 NOTYPE  GLOBAL DEFAULT  ABS __bss_start
     8: 0000033c     0 FUNC    GLOBAL DEFAULT    9 _init
     9: 000004b8     0 FUNC    GLOBAL DEFAULT   12 _fini
    10: 0000045c    28 FUNC    GLOBAL DEFAULT   11 sharedLib1Function
milan@milan$
```

Figure 12-23. *Using readelf to list dynamic symbols*

Examining the Dynamic Section

Running readelf -d examines the dynamic section (useful for finding DT_RPATH and/or DT_RUNPATH settings), as shown in Figure 12-24.

```
milan@milan$ readelf -d demoApp

Dynamic section at offset 0x1dd0 contains 28 entries:
  Tag        Type                         Name/Value
 0x0000000000000001 (NEEDED)             Shared library: [libpthread.so.0]
 0x0000000000000001 (NEEDED)             Shared library: [libdl.so.2]
 0x0000000000000001 (NEEDED)             Shared library: [libdynamiclinkingdemo.so]
 0x0000000000000001 (NEEDED)             Shared library: [libstdc++.so.6]
 0x0000000000000001 (NEEDED)             Shared library: [libm.so.6]
 0x0000000000000001 (NEEDED)             Shared library: [libgcc_s.so.1]
 0x0000000000000001 (NEEDED)             Shared library: [libc.so.6]
 0x000000000000001d (RUNPATH)            Library runpath: [../deploy:./deploy]
 0x000000000000000c (INIT)               0x4005d8
 0x000000000000000d (FINI)               0x400808
 0x0000000000000004 (HASH)               0x3ff4d0
 0x000000006ffffef5 (GNU_HASH)           0x3ff490
 0x0000000000000005 (STRTAB)             0x3ff270
 0x0000000000000006 (SYMTAB)             0x3ff388
 0x000000000000000a (STRSZ)              277 (bytes)
 0x000000000000000b (SYMENT)             24 (bytes)
 0x0000000000000015 (DEBUG)              0x0
 0x0000000000000003 (PLTGOT)             0x600fe8
 0x0000000000000002 (PLTRELSZ)           72 (bytes)
 0x0000000000000014 (PLTREL)             RELA
 0x0000000000000017 (JMPREL)             0x400590
 0x0000000000000007 (RELA)               0x400578
 0x0000000000000008 (RELASZ)             24 (bytes)
 0x0000000000000009 (RELAENT)            24 (bytes)
 0x000000006ffffffe (VERNEED)            0x400538
 0x000000006fffffff (VERNEEDNUM)         2
 0x000000006ffffff0 (VERSYM)             0x400522
 0x0000000000000000 (NULL)               0x0
milan@milan$
```

Figure 12-24. *Using readelf to display the dynamic section*

Examining the Relocation Section

Running readelf -r examines the relocation section, as shown in Figure 12-25.

```
milan@milan$ readelf -r libmreloc.so

Relocation section '.rel.dyn' at offset 0x2dc contains 7 entries:
 Offset     Info    Type              Sym.Value  Sym. Name
00002008  00000008 R_386_RELATIVE
00000450  00000401 R_386_32            0000200c   myglob
00000458  00000401 R_386_32            0000200c   myglob
0000045d  00000401 R_386_32            0000200c   myglob
00001fe8  00000106 R_386_GLOB_DAT      00000000   __cxa_finalize
00001fec  00000206 R_386_GLOB_DAT      00000000   __gmon_start__
00001ff0  00000306 R_386_GLOB_DAT      00000000   _Jv_RegisterClasses

Relocation section '.rel.plt' at offset 0x314 contains 2 entries:
 Offset     Info    Type              Sym.Value  Sym. Name
00002000  00000107 R_386_JUMP_SLOT     00000000   __cxa_finalize
00002004  00000207 R_386_JUMP_SLOT     00000000   __gmon_start__
milan@milan$
```

Figure 12-25. *Using readelf to list relocation (.rel.dyn) section*

Examining the Data Section

Running readelf -x provides the hex dump of the values carried by the section. In Figure 12-26, it is the .got section.

```
milan@milan$ readelf -x .got driver

Hex dump of section '.got':
  0x08049ff0 00000000                            ....

milan@milan$
```

Figure 12-26. *Using readelf to provide a hex dump of a section (the .got section in this example)*

Listing and Examining Segments

Running readelf --segments displays information about the ELF binary segments (Figure 12-27).

```
milan@milan$ readelf --segments libmreloc.so

Elf file type is DYN (Shared object file)
Entry point 0x390
There are 7 program headers, starting at offset 52

Program Headers:
  Type           Offset   VirtAddr   PhysAddr   FileSiz MemSiz  Flg Align
  LOAD           0x000000 0x00000000 0x00000000 0x00540 0x00540 R E 0x1000
  LOAD           0x000f0c 0x00001f0c 0x00001f0c 0x00104 0x0010c RW  0x1000
  DYNAMIC        0x000f20 0x00001f20 0x00001f20 0x000c8 0x000c8 RW  0x4
  NOTE           0x000114 0x00000114 0x00000114 0x00024 0x00024 R   0x4
  GNU_EH_FRAME   0x0004c4 0x000004c4 0x000004c4 0x0001c 0x0001c R   0x4
  GNU_STACK      0x000000 0x00000000 0x00000000 0x00000 0x00000 RW  0x4
  GNU_RELRO      0x000f0c 0x00001f0c 0x00001f0c 0x000f4 0x000f4 R   0x1

 Section to Segment mapping:
  Segment Sections...
   00     .note.gnu.build-id .gnu.hash .dynsym .dynstr .gnu.version .gnu.version_r
          .rel.dyn .rel.plt .init .plt .text .fini .eh_frame_hdr .eh_frame
   01     .ctors .dtors .jcr .dynamic .got .got.plt .data .bss
   02     .dynamic
   03     .note.gnu.build-id
   04     .eh_frame_hdr
   05
   06     .ctors .dtors .jcr .dynamic .got
milan@milan$
```

Figure 12-27. *Using readelf to examine segments*

Detecting the Debug Build

The readelf command has very good support for displaying all kind of debugging-specific information contained in the binary (Figure 12-28).

```
READELF(1)                    GNU Development Tools                    READELF(1)

NAME
       readelf - Displays information about ELF files.

SYNOPSIS
       readelf [-a|--all]
               [-h|--file-header]
               [-l|--program-headers|--segments]
               [-S|--section-headers|--sections]
               [-g|--section-groups]
               [-t|--section-details]
               [-e|--headers]
               [-s|--syms|--symbols]
               [--dyn-syms]
               [-n|--notes]
               [-r|--relocs]
               [-u|--unwind]
               [-d|--dynamic]
               [-V|--version-info]
               [-A|--arch-specific]
               [-D|--use-dynamic]
               [-x <number or name>|--hex-dump=<number or name>]
               [-p <number or name>|--string-dump=<number or name>]
               [-R <number or name>|--relocated-dump=<number or name>]
               [-c|--archive-index]
               [-w[lLiaprmfFsoRt]|
                --debug-dump[=rawline,=decodedline,=info,=abbrev,=pubnames,=aran
ges,=macro,=frames,=frames-interp,=str,=loc,=Ranges,=pubtypes,=trace_info,=trace
_abbrev,=trace_aranges,=gdb_index]]
```

Figure 12-28. Readelf provides the option to examine binary file debug information

To quickly determine whether the binary was built for debug or not, in the case of a debug build, the output of running readelf --debug-dump with any of available options will be comprised of a number of lines printed on stdout. On the contrary, if the binary was not built for debugging, the output will be an empty line. One of the quick and practical ways of limiting the output spew in the case when the binary contains the debugging information is to pipe the readelf output to the wc command:

$ *readelf --debug-dump=line* <binary file path>| wc -l

Alternatively, the following simple script may be used to display the readelf's findings in plain and simple text form. It requires that the path to the binary file be passed as an input argument.

file: isDebugVersion.sh
if *readelf --debug-dump=line* $1 > /dev/null; then echo "$1 is built for debug"; fi

Deployment Phase Tools

After you successfully build your binary files and start thinking about the details of the deploying stage, utilities such as chrpath, patchelf, strip, and ldconfig may come handy.

chrpath

The chrpath command-line utility program (http://linux.die.net/man/1/chrpath) is used to modify the rpath (DT_RPATH field) of the ELF binaries. The basic concept behind the runpath field is described in Chapter 7, under the "Linux Runtime Library Location Rules" section.

The following details illustrate the use (Figure 12-29) as well as some of the limitations (Figure 12-30) of chrpath:

- It can be used to modify DT_RPATH within its original string length.

- It can be used to delete the existing DT_RPATH field.

  However, be cautious!

- If the DT_RPATH string is initially empty, it cannot be substituted with new non-empty string.

- It can be used to convert DT_RPATH to DT_RUNPATH.

- It cannot substitute the existing DT_RPATH string with the longer string.

```
milan@milan$ readelf -d demo_rpath | grep RPATH
 0x0000000f (RPATH)                      Library rpath: [/home/milan/Desktop/Test]
milan@milan$ chrpath -r /home/john/Desktop/Test ./demo_rpath
./demo_rpath: RPATH=/home/milan/Desktop/Test
./demo_rpath: new RPATH: /home/john/Desktop/Test       1) can modify the existing RPATH within the
milan@milan$ readelf -d demo_rpath | grep RPATH           original string length
 0x0000000f (RPATH)                      Library rpath: [/home/john/Desktop/Test]
milan@milan$ chrpath -c ./demo_rpath
./demo_rpath: RPATH converted to RUNPATH              2) can convert RPATH to RUNPATH
./demo_rpath: RUNPATH=/home/john/Desktop/Test
milan@milan$ readelf -d demo_rpath | grep PATH
 0x0000001d (RUNPATH)                    Library runpath: [/home/john/Desktop/Test]
milan@milan$                                          3) can't make RPATH string longer
milan@milan$ chrpath -r /exceptionally/long/new/rpath/for/demo/purposes ./demo_rpath
./demo_rpath: RUNPATH=/home/john/Desktop/Test
new rpath '/exceptionally/long/new/rpath/for/demo/purposes' too large; maximum length 24
milan@milan$
```

```
milan@milan$ readelf -d demo_rpath | grep PATH
 0x0000000f (RPATH)                      Library rpath: [/home/milan/Desktop/Test]
milan@milan$ chrpath -d demo_rpath                    chrpath can delete the existing RPATH
milan@milan$ readelf -d demo_rpath | grep PATH
milan@milan$
```

Figure 12-29. *Using the chrpath utility to modify RPATH*

```
milan@milan$ ls -alg
total 12
drwxrwxr-x 2 milan 4096 Apr 28 12:30 .
drwxr-xr-x 4 milan 4096 Apr 28 12:14 ..
-rw-rw-r-- 1 milan   95 Apr 28 12:15 main.cpp
milan@milan$ gcc main.cpp -o demo_no_rpath_set_initially
milan@milan$ readelf -d ./demo_no_rpath_set_initially

Dynamic section at offset 0xf28 contains 20 entries:
  Tag        Type                     Name/Value
 0x00000001 (NEEDED)                  Shared library: [libc.so.6]
 0x0000000c (INIT)                    0x80482b0
 0x0000000d (FINI)                    0x804849c
 0x6ffffef5 (GNU_HASH)                0x80481ac
 0x00000005 (STRTAB)                  0x804821c
 0x00000006 (SYMTAB)                  0x80481cc
 0x0000000a (STRSZ)                   74 (bytes)
 0x0000000b (SYMENT)                  16 (bytes)
 0x00000015 (DEBUG)                   0x0
 0x00000003 (PLTGOT)                  0x8049ff4
 0x00000002 (PLTRELSZ)                24 (bytes)
 0x00000014 (PLTREL)                  REL         if the RPATH string is empty
 0x00000017 (JMPREL)                  0x8048298   (nonexistent) chrpath can not
 0x00000011 (REL)                     0x8048290   replace it with a new non-empty
 0x00000012 (RELSZ)                   8 (bytes)   value
 0x00000013 (RELENT)                  8 (bytes)
 0x6ffffffe (VERNEED)                 0x8048270
 0x6fffffff (VERNEEDNUM)              1
 0x6ffffff0 (VERSYM)                  0x8048266
 0x00000000 (NULL)                    0x0
milan@milan$ chrpath -c /home/milan/Desktop/ ./demo_no_rpath_set_initially
open: Is a directory
elf_open: Is a directory
milan@milan$
```

Figure 12-30. *Limitations of the chrpath utility*

patchelf

The useful patchelf (http://nixos.org/patchelf.html) command-line utility program is currently not part of the standard repositories, but it is possible to build it from the source tarball. Simple, basic documentation is also available.

This utility can be used to set and modify the runpath (DT_RUNPATH field) of the ELF binaries. The basic concept behind the runpath field is described in the Chapter 7, under the "Linux Runtime Library Location Rules" section.

The simplest way of setting up the runpath is to issue a command like this one:

```
$ patchelf --set-rpath <one or more paths> <executable>
                        ^
                        |
                     multiple paths can be defined,
                     separated by a colon (:)
```

The capabilities of patchelf of modifying the DT_RUNPATH field far exceed the chrpath capabilities of modifying the DT_RPATH field, as it can modify the string value of DT_RUNPATH in any way imaginable (substituting with a shorter or longer string, inserting multiple paths, erasing, etc.).

strip

The strip command-line utility program (http://linux.die.net/man/1/strip) can be used to eliminate all the library symbols that are not needed in the process of dynamic loading. An illustration of the strip effects was demonstrated in Chapter 7, under the "Exporting the Linux dynamic library symbols" section.

ldconfig

In Chapter 7 (devoted to the Linux runtime library location rules), I indicated that one of the ways (albeit not the highest priority) to specify the paths where the loader should look for libraries at runtime is through the use of ldconfig cache.

The ldconfig command-line utility program (http://linux.die.net/man/8/ldconfig) is typically executed as the very last step of a package installation procedure. When a path containing the shared library is passed as input argument to ldconfig, it searches the path for the shared libraries, and updates the set of files it uses for bookkeeping:

- The file /etc/ld.so.conf containing the list of folders it standardly scans

- The file /etc/ld.so.cache file, which contain the ASCII textual list of all libraries found throughout the scans of variety of paths that were passed as input argument

Runtime Analysis Tools

The analyses of runtime issues may benefit by using the tools such as strace, addr2line, and especially GNU debugger (gdb).

strace

The strace (http://linux.die.net/man/1/strace) command-line utility program tracks down the system calls made by the process as well as the signals received by the process. It can be helpful in figuring out the runtime dependencies (i.e., not only load-time dependencies for which the ldd command is suitable). Figure 12-31 illustrates the typical strace output.

```
milan@milan$ strace
execve("./driver", ["./driver"], [/* 36 vars */]) = 0
brk(0)                                    = 0x80cf000
access("/etc/ld.so.nohwcap", F_OK)       = -1 ENOENT (No such file or directory)
mmap2(NULL, 8192, PROT_READ|PROT_WRITE, MAP_PRIVATE|MAP_ANONYMOUS, -1, 0) = 0xb773a000
access("/etc/ld.so.preload", R_OK)       = -1 ENOENT (No such file or directory)
open("../sharedLib/tls/i686/sse2/cmov/libmreloc.so", O_RDONLY|O_CLOEXEC) = -1 ENOENT (No such file or directory)
open("../sharedLib/tls/i686/sse2/libmreloc.so", O_RDONLY|O_CLOEXEC) = -1 ENOENT (No such file or directory)
open("../sharedLib/tls/i686/cmov/libmreloc.so", O_RDONLY|O_CLOEXEC) = -1 ENOENT (No such file or directory)
open("../sharedLib/tls/i686/libmreloc.so", O_RDONLY|O_CLOEXEC) = -1 ENOENT (No such file or directory)
open("../sharedLib/tls/sse2/cmov/libmreloc.so", O_RDONLY|O_CLOEXEC) = -1 ENOENT (No such file or directory)
open("../sharedLib/tls/sse2/libmreloc.so", O_RDONLY|O_CLOEXEC) = -1 ENOENT (No such file or directory)
open("../sharedLib/tls/cmov/libmreloc.so", O_RDONLY|O_CLOEXEC) = -1 ENOENT (No such file or directory)
open("../sharedLib/tls/libmreloc.so", O_RDONLY|O_CLOEXEC) = -1 ENOENT (No such file or directory)
open("../sharedLib/i686/sse2/cmov/libmreloc.so", O_RDONLY|O_CLOEXEC) = -1 ENOENT (No such file or directory)
open("../sharedLib/i686/sse2/libmreloc.so", O_RDONLY|O_CLOEXEC) = -1 ENOENT (No such file or directory)
open("../sharedLib/i686/cmov/libmreloc.so", O_RDONLY|O_CLOEXEC) = -1 ENOENT (No such file or directory)
open("../sharedLib/i686/libmreloc.so", O_RDONLY|O_CLOEXEC) = -1 ENOENT (No such file or directory)
open("../sharedLib/sse2/cmov/libmreloc.so", O_RDONLY|O_CLOEXEC) = -1 ENOENT (No such file or directory)
open("../sharedLib/sse2/libmreloc.so", O_RDONLY|O_CLOEXEC) = -1 ENOENT (No such file or directory)
open("../sharedLib/cmov/libmreloc.so", O_RDONLY|O_CLOEXEC) = -1 ENOENT (No such file or directory)
open("../sharedLib/libmreloc.so", O_RDONLY|O_CLOEXEC) = 3
read(3, "\177ELF\1\1\1\0\0\0\0\0\0\0\0\0\3\0\3\0\1\0\0\0\260\3\0\0004\0\0\0"..., 512) = 512
fstat64(3, {st_mode=S_IFREG|0775, st_size=7727, ...}) = 0
getcwd("/home/milan/Desktop/Test/loadTimeRelocation/example2/driverApp", 128) = 63
mmap2(NULL, 8216, PROT_READ|PROT_EXEC, MAP_PRIVATE|MAP_DENYWRITE, 3, 0) = 0xb7737000
mmap2(0xb7738000, 8192, PROT_READ|PROT_WRITE, MAP_PRIVATE|MAP_FIXED|MAP_DENYWRITE, 3, 0) = 0xb7738000
close(3)                                  = 0
open("../sharedLib/tls/i686/sse2/cmov/libc.so.6", O_RDONLY|O_CLOEXEC) = -1 ENOENT (No such file or directory)
open("../sharedLib/tls/i686/sse2/libc.so.6", O_RDONLY|O_CLOEXEC) = -1 ENOENT (No such file or directory)
open("../sharedLib/tls/i686/cmov/libc.so.6", O_RDONLY|O_CLOEXEC) = -1 ENOENT (No such file or directory)
open("../sharedLib/tls/i686/libc.so.6", O_RDONLY|O_CLOEXEC) = -1 ENOENT (No such file or directory)
open("../sharedLib/tls/sse2/cmov/libc.so.6", O_RDONLY|O_CLOEXEC) = -1 ENOENT (No such file or directory)
open("../sharedLib/tls/sse2/libc.so.6", O_RDONLY|O_CLOEXEC) = -1 ENOENT (No such file or directory)
open("../sharedLib/tls/cmov/libc.so.6", O_RDONLY|O_CLOEXEC) = -1 ENOENT (No such file or directory)
open("../sharedLib/tls/libc.so.6", O_RDONLY|O_CLOEXEC) = -1 ENOENT (No such file or directory)
open("../sharedLib/i686/sse2/cmov/libc.so.6", O_RDONLY|O_CLOEXEC) = -1 ENOENT (No such file or directory)
open("../sharedLib/i686/sse2/libc.so.6", O_RDONLY|O_CLOEXEC) = -1 ENOENT (No such file or directory)
open("../sharedLib/i686/cmov/libc.so.6", O_RDONLY|O_CLOEXEC) = -1 ENOENT (No such file or directory)
open("../sharedLib/i686/libc.so.6", O_RDONLY|O_CLOEXEC) = -1 ENOENT (No such file or directory)
open("../sharedLib/sse2/cmov/libc.so.6", O_RDONLY|O_CLOEXEC) = -1 ENOENT (No such file or directory)
open("../sharedLib/sse2/libc.so.6", O_RDONLY|O_CLOEXEC) = -1 ENOENT (No such file or directory)
open("../sharedLib/cmov/libc.so.6", O_RDONLY|O_CLOEXEC) = -1 ENOENT (No such file or directory)
open("../sharedLib/libc.so.6", O_RDONLY|O_CLOEXEC) = -1 ENOENT (No such file or directory)
open("/etc/ld.so.cache", O_RDONLY|O_CLOEXEC) = 3
fstat64(3, {st_mode=S_IFREG|0644, st_size=70505, ...}) = 0
mmap2(NULL, 70505, PROT_READ, MAP_PRIVATE, 3, 0) = 0xb7725000
```

Figure 12-31. *Using the strace utility*

addr2line

The addr2line (http://linux.die.net/man/1/addr2line) command-line utility program can be used to convert the runtime address into information about the source file and the line number corresponding to the address.

If (and if only) the binary is built for debug (by passing the -g -O0 compiler flags), using this command may be very helpful when analyzing crash information in which the program counter address where the crash happened is printed on the terminal screen as something like this:

```
#00 pc 0000d8cc6 /usr/mylibs/libxyz.so
```

Running the addr2line on such console output

```
$ addr2line -C -f -e /usr/mylibs/libxyz.so 0000d8cc6
```

will result with output that may look like this:

```
/projects/mylib/src/mylib.c: 45
```

gdb (GNU Debugger)

The legendary GNU debugger tool known as gdb can be used to perform the runtime code disassembling. The advantage of disassembling the code at runtime is that all the addresses have already been resolved by the loader, and the addresses are for the most part final.

The following gdb commands can be useful during the runtime code disassembling:

- ***set disassembly-flavor*** *<intel | att>*

- ***disassemble*** <function name>

The following two flags may come handy when invoking the disassemble command:

- The /r flag requires that the assembler instructions be additionally shown in hexadecimal notation (Figure 12-32).

```
(gdb) set disassembly-flavor intel
(gdb) disassemble /r main
Dump of assembler code for function main:
   0x08048875 <+0>:     55            push    ebp
   0x08048876 <+1>:     89 e5         mov     ebp,esp
   0x08048878 <+3>:     83 e4 f0      and     esp,0xfffffff0
   0x0804887b <+6>:     83 ec 20      sub     esp,0x20
   0x0804887e <+9>:     c7 44 24 14 00 00 00 00  mov   DWORD PTR [esp+0x14],0x0
   0x08048886 <+17>:    c7 44 24 04 00 00 00 00  mov   DWORD PTR [esp+0x4],0x0
   0x0804888e <+25>:    c7 04 24 5f 87 04 08     mov   DWORD PTR [esp],0x804875f
   0x08048895 <+32>:    e8 a6 fc ff ff  call    0x8048540 <dl_iterate_phdr@plt>
   0x0804889a <+37>:    a1 30 a0 04 08  mov     eax,ds:0x804a030
   0x0804889f <+42>:    83 c0 01      add     eax,0x1
   0x080488a2 <+45>:    89 44 24 18   mov     DWORD PTR [esp+0x18],eax
   0x080488a6 <+49>:    8b 45 08      mov     eax,DWORD PTR [ebp+0x8]
   0x080488a9 <+52>:    89 44 24 04   mov     DWORD PTR [esp+0x4],eax
   0x080488ad <+56>:    8b 44 24 18   mov     eax,DWORD PTR [esp+0x18]
   0x080488b1 <+60>:    89 04 24      mov     DWORD PTR [esp],eax
   0x080488b4 <+63>:    e8 a7 fc ff ff  call    0x8048560 <initialize@plt>
   0x080488b9 <+68>:    89 44 24 14   mov     DWORD PTR [esp+0x14],eax
   0x080488bd <+72>:    b8 f8 8b 04 08  mov     eax,0x8048bf8
   0x080488c2 <+77>:    8b 54 24 14   mov     edx,DWORD PTR [esp+0x14]
   0x080488c6 <+81>:    89 54 24 0c   mov     DWORD PTR [esp+0xc],edx
```

Figure 12-32. *Using gdb to show the disassembled code combined with hex values of instructions*

- The /m flag intersperses the assembler instructions with the C/C++ lines of code (if available), as shown in Figure 12-33.

```
(gdb) disassemble /m main
Dump of assembler code for function main:
117     {
   0x08048875 <+0>:      push    ebp
   0x08048876 <+1>:      mov     ebp,esp
   0x08048878 <+3>:      and     esp,0xfffffff0
   0x0804887b <+6>:      sub     esp,0x20

118            int t = 0;
   0x0804887e <+9>:      mov     DWORD PTR [esp+0x14],0x0

119        dl_iterate_phdr(header_handler, NULL);
   0x08048886 <+17>:     mov     DWORD PTR [esp+0x4],0x0
   0x0804888e <+25>:     mov     DWORD PTR [esp],0x804875f
   0x08048895 <+32>:     call    0x8048540 <dl_iterate_phdr@plt>

120
121        int first = shlibNonStaticAccessedAsExternVariable + 1;
   0x0804889a <+37>:     mov     eax,ds:0x804a030
   0x0804889f <+42>:     add     eax,0x1
   0x080488a2 <+45>:     mov     DWORD PTR [esp+0x18],eax

122            t = initialize(first, argc);
   0x080488a6 <+49>:     mov     eax,DWORD PTR [ebp+0x8]
```

Figure 12-33. *Interspersed (assembly and source code) disassembly flavor*

To combine these two flags, type them together (i.e., /rm) instead of separately (i.e., /r /m), as shown in Figure 12-34.

```
(gdb) disassemble /mr main
Dump of assembler code for function main:
117     {
   0x08048875 <+0>:      55          push    ebp
   0x08048876 <+1>:      89 e5       mov     ebp,esp
   0x08048878 <+3>:      83 e4 f0        and     esp,0xfffffff0
   0x0804887b <+6>:      83 ec 20        sub     esp,0x20

118            int t = 0;
   0x0804887e <+9>:      c7 44 24 14 00 00 00 00 mov     DWORD PTR [esp+0x14],0x0

119        dl_iterate_phdr(header_handler, NULL);
   0x08048886 <+17>:     c7 44 24 04 00 00 00 00 mov     DWORD PTR [esp+0x4],0x0
   0x0804888e <+25>:     c7 04 24 5f 87 04 08    mov     DWORD PTR [esp],0x804875f
   0x08048895 <+32>:     e8 a6 fc ff ff  call    0x8048540 <dl_iterate_phdr@plt>

120
121        int first = shlibNonStaticAccessedAsExternVariable + 1;
   0x0804889a <+37>:     a1 30 a0 04 08  mov     eax,ds:0x804a030
   0x0804889f <+42>:     83 c0 01        add     eax,0x1
   0x080488a2 <+45>:     89 44 24 18     mov     DWORD PTR [esp+0x18],eax
```

Figure 12-34. *Combining /r and /m disassembly flags*

Static Library Tools

The vast majority of tasks related to the static libraries can be performed by the archiver ar utility. By using ar, you may not only combine the object files into the static library, but also list its contents, remove individual object files, or replace them with the newer version.

ar

The following simple example illustrates the usual stages of using the ar tool. The demo project is comprised of four source files (first.c, second.c, third.c, and fourth.c) and one export header file that can be used by the client binaries (shown in following five examples),

first.c

```
#include "mystaticlibexports.h"

int first_function(int x)
{
        return (x+1);
}
```

second.c

```
#include "mystaticlibexports.h"

int fourth_function(int x)
{
        return (x+4);
}
```

third.c

```
#include "mystaticlibexports.h"

int second_function(int x)
{
        return (x+2);
}
```

fourth.c

```
#include "mystaticlibexports.h"

int third_function(int x)
{
        return (x+3);
}
```

mystaticlibexports.h

```
#pragma once
int first_function(int x);
int second_function(int x);
int third_function(int x);
int fourth_function(int x);
```

Let's assume that you have the object files created by compiling each of the source files:

```
$ gcc -Wall -c first.c second.c third.c fourth.c
```

The following screen snapshots illustrate the various stages of dealing with the static library.

Creating the Static Library

Running `ar -rcs <library name> <list of object files>` combines the specified object files into the static library (Figure 12-35).

```
milan@milan$ ar -rcs libmystaticlib.a first.o second.o third.o fourth.o
milan@milan$ ls -alg
total 48
drwxrwxr-x 2 milan 4096 Dec 25 11:37 .
drwxrwxr-x 5 milan 4096 Dec 25 10:48 ..
-rw-rw-r-- 1 milan   78 Dec 25 10:36 first.c
-rw-rw-r-- 1 milan  864 Dec 25 11:35 first.o
-rw-rw-r-- 1 milan   79 Dec 25 10:36 fourth.c
-rw-rw-r-- 1 milan  868 Dec 25 11:35 fourth.o
-rw-rw-r-- 1 milan 3854 Dec 25 11:37 libmystaticlib.a
-rw-rw-r-- 1 milan  124 Dec 25 10:37 mystaticlibexports.h
-rw-rw-r-- 1 milan   79 Dec 25 10:35 second.c
-rw-rw-r-- 1 milan  868 Dec 25 11:35 second.o
-rw-rw-r-- 1 milan   78 Dec 25 10:35 third.c
-rw-rw-r-- 1 milan  864 Dec 25 11:35 third.o
milan@milan$ file libmystaticlib.a
libmystaticlib.a: current ar archive
```

Figure 12-35. *Using ar to combine object files to static library*

Listing the Static Library Object Files

Running `ar -t <library name>` prints out the list of the object files carried by the static library (Figure 12-36).

```
milan@milan$ ar -t libmystaticlib.a
first.o
second.o
third.o
fourth.o
milan@milan$
```

Figure 12-36. *Using ar to print out the list of static library's object files*

Deleting an Object File from the Static Library

Let's say that you want to modify the file first.c (to fix a bug, or simply to add extra feature) and for the time being you don't want your static library to carry the first.o object file. The way to delete the object file from the static library is to run ar -d <library name> <object file to remove> (Figure 12-37).

```
milan@milan$ ar -t libmystaticlib.a
first.o
second.o
third.o
fourth.o
milan@milan$ ar -d libmystaticlib.a first.o
milan@milan$ ar -t libmystaticlib.a
second.o
third.o
fourth.o
milan@milan$
```

Figure 12-37. Using ar to delete an object file from static library

Adding the New Object File to the Static Library

Let's say that you are happy with the changes you made in the file first.c and you have recompiled it. Now you want to put the newly created object file first.o back into the static library. Running ar -r <library name> <object file to append> basically appends your new object file to the static library (Figure 12-38).

```
milan@milan$ cat first.c
#include <stdio.h>
#include "mystaticlibexports.h"

int first_function(int x)
{
    printf("%s\n", __FUNCTION__);
    return (x+1);
}
milan@milan$ gcc -Wall -I../staticLib -c first.c
milan@milan$ ar -r libmystaticlib.a first.o
milan@milan$ ar -t libmystaticlib.a
second.o
third.o
fourth.o
first.o
milan@milan$
```

Figure 12-38. Using ar to add new object file to static library

Please notice that the order in which the object files reside in the static library has been changed. The new file has been effectively appended to the archive.

Restoring the Order of Object Files

If you insist on your object files appearing in the original order that existed prior to your code changes, you may correct it. Running ar -m -b <object file before> <library name> <object file to move> accomplishes the task (Figure 12-39).

```
milan@milan$ ar -t libmystaticlib.a
second.o
third.o
fourth.o
first.o
milan@milan$ ar -m -b second.o libmystaticlib.a first.o
milan@milan$ ar -t libmystaticlib.a
first.o
second.o
third.o
fourth.o
milan@milan$
```

Figure 12-39. Using ar to restore the order of object files within the static library

CHAPTER 13

■ ■ ■

Linux How To's

The previous chapter provided a review of the useful analysis utilities available in Linux, so now is a good moment to provide an alternative view of the same topic. This time the focus will not be on the utilities per se, but rather on showing how some of the most frequently performed tasks can be completed.

There is typically more than one way to complete an analysis task. For each of the tasks described in this chapter, alternative ways of completing the task will be provided.

Debugging the Linking

Probably the most powerful aid in debugging the linking stage is the use of the LD_DEBUG environment variable (Figure 13-1). It is suitable for testing not only the build process but also the dynamic library loading at runtime.

```
milan@milan$ LD_DEBUG=help cat
Valid options for the LD_DEBUG environment variable are:

  libs        display library search paths
  reloc       display relocation processing
  files       display progress for input file
  symbols     display symbol table processing
  bindings    display information about symbol binding
  versions    display version dependencies
  scopes      display scope information
  all         all previous options combined
  statistics  display relocation statistics
  unused      determined unused DSOs
  help        display this help message and exit

To direct the debugging output into a file instead of standard output
a filename can be specified using the LD_DEBUG_OUTPUT environment variable.
milan@milan$
```

Figure 13-1. *Using the LD_DEBUG environment variable to debug linking*

The operating system supports a predetermined set of values to which LD_DEBUG may be set before running the desired operation (build or execution). The way to have them listed is to type

```
$ LD_DEBUG=help cat
```

As with any other environment variable, there are several ways to set the value of LD_DEBUG:

- Immediately, on the same line from which the linker is invoked

- Once for the lifetime of terminal shell

  ```
  $ export LD_DEBUG=<chosen_option>
  ```

  which can be reversed by

  ```
  $ unset LD_DEBUG
  ```

- From within the shell profile (such as .bashrc) file, setting it for each and every terminal session. Unless your daily job is to test the linking process, this option is probably not the most optimal one.

Determining the Binary File Type

There are a few simple ways to determine the binary type:

- The file utility (among the wide variety of file types it can handle) provides probably the simplest, quickest, and most elegant way to determine the nature of binary file.

- readelf ELF header analysis provides, among other details, information about the binary file type. Running

  ```
  $ readelf -h <path-of-binary> | grep Type
  ```

  will display one of the following choices:

  - EXEC (executable file)

  - DYN (shared object file)

  - REL (relocatable file)

  In the case of static libraries, the REL output will appear once for each of the object files carried by the library.

- objdump EFL header analysis may provide similar analysis with a bit less detailed report. The output of this command

  ```
  $ objdump -f <path-of-binary>
  ```

  will have a line containing one of the following values:

  - EXEC_P (executable file)

  - DYNAMIC (shared object file)

  - No type indicated, in the case of an object file

  In the case of a static library, an object file will appear once for each of the object files carried by the library.

Determining the Binary File Entry Point

Determining the binary file entry point is a task that varies in complexity from the very simple (in the case of executable files) to the somewhat more involved (determining the entry point of the dynamic library at runtime), both of which will be illustrated in this section.

Determining the Executable Entry Point

The entry point of executable (i.e., the address of the first instruction in the program memory map) can be determined by either

- readelf ELF header analysis, which provides, among other details, information about the binary file type. Running

  ```
  $ readelf -h <path-of-binary> | grep Entry
  ```

  will display a line looking somewhat like this:

  ```
  Entry point address:              0x<address>
  ```

- objdump EFL header analysis, which may provide the similar analysis with a bit less detailed report. The output of this command

  ```
  $ objdump -f <path-of-binary> | grep start
  ```

  will displ.ne looking somewhat like this:

  ```
  start address 0x<address>
  ```

Determining the Dynamic Library Entry Point

When the entry point of a dynamic library is looked for, the investigation is not as straightforward. Even though it is possible to use one of the previously described methods, the provided information (typically a low-valued hex number, such as 0x390) is not particularly useful. Given the fact that the dynamic library is mapped into the client binary process memory map, the library's true entry point may be determined only at runtime.

Probably the simplest way is to run the executable that loads the dynamic library in the gnu debugger. If the LD_DEBUG environment variable is set, the information about the loaded library will be printed out. All you need to do is to set the break point on the main() function. This symbol is very likely to exist regardless of whether the executable was built for debugging or not.

In cases when the dynamic library is linked in a statically aware fashion, by the time the program execution reaches the breakpoint there, the loading process will already be completed.

In cases of runtime dynamic loading, probably the easiest approach is to redirect the massive screen printout to the file for visual inspection later on.

Figure 13-2 illustrates the method that relies on the LD_DEBUG variable.

```
milan@milan$ LD_DEBUG=files gdb -q ./driver
     3226:
     3226:        file=libreadline.so.6 [0];  needed by gdb [0]
     3226:        file=libreadline.so.6 [0];  generating link map
     3226:          dynamic: 0xb775bb88  base: 0xb7726000  size: 0x00039de4
     3226:            entry: 0xb7730ef0  phdr: 0xb7726034  phnum:        7
     3226:
                                      o
                                      o
                                      o
Reading symbols from /home/milan/driverApp/driver...done.
(gdb) b main
Breakpoint 1 at 0x804864f: file driver.c, line 28.
(gdb) r
Starting program: /home/milan/driverApp/driver
     3229:
     3229:        file=libtinfo.so.5 [0];  needed by /bin/bash [0]
     3229:        file=libtinfo.so.5 [0];  generating link map

                                      o
                                      o
                                      o

     3229:
     3229:        file=libmreloc.so [0];  needed by /home/milan/driverApp/driver [0]
     3229:        file=libmreloc.so [0];  generating link map
     3229:          dynamic: 0xb7fd9f20  base: 0xb7fd8000  size: 0x00002018
     3229:            entry: 0xb7fd8390  phdr: 0xb7fd8034  phnum:        7
     3229:
     3229:
```

○
○ Please notice that
○ entry - base = 0x390,
  which is the value read out
  by readelf from the library binary

```
Breakpoint 1, main (argc=1, argv=0xbffff344) at driver.c:28
28              dl_iterate_phdr(header_handler, NULL);
(gdb) set disassembly-flavor intel
(gdb) disassemble 0xb7fd8390
Dump of assembler code for function __do_global_dtors_aux:
   0xb7fd8390 <+0>:     push   ebp
   0xb7fd8391 <+1>:     mov    ebp,esp
   0xb7fd8393 <+3>:     push   esi
   0xb7fd8394 <+4>:     push   ebx
   0xb7fd8395 <+5>:     call   0xb7fd8447 < i686.get_pc_thunk.bx>
```

1) before running the debugger, activate linker debugging by setting the LD_DEBUG environment variable (choose the 'files' option.

2) Start the debugger

3) set the breakpoint at a convention place (such as 'main' symbol).

4) Run the process as it will load all the required libraries (assuming run time location rules have been satisfied).

5) active LD_DEBUG will print the run-time loading details of library we are interested in

6) when breakpoint gets hit try disassembling the code around the address being reported as run-time entry point address.

If everything is OK, gdb will report that this address is also a function entry point.

Figure 13-2. *Determining the dynamic library entry point at runtimeList Symbols*

List Symbols

The following approaches may be followed when trying to list the symbols of executables and libraries:

- nm utility

- readelf utility

  In particular,

  - A list of all visible symbols may be obtained by running

    ```
    $ readelf --symbols <path-to-binary>
    ```

- A list of only the symbols exported for dynamic linking purposes may be obtained by running

  ```
  $ readelf --dyn-syms <path-to-binary>
  ```

- objdump utility

  In particular,

  - A list of all visible symbols may be obtained by running

    ```
    $ objdump -t <path-to-binary>
    ```

  - A list of only the symbols exported for dynamic linking purposes may be obtained by running

    ```
    $ objdump -T <path-to-binary>
    ```

List and Examine Sections

There are several ways of obtaining the information about the binary sections. The quickest and most rudimentary insight can be obtained by running the size command. For a more structured and more detailed insight, you can typically rely on tools like objdump and/or readelf, the latter being specialized strictly in the ELF binary format. Typically, the mandatory first step is to list all the sections present in the binary file. Once such insight is obtained, the content of a specific segment is examined in detail.

Listing the Available Sections

The list of sections of the ELF binary file can be obtained by one of the following methods:

- readelf utility

  ```
  $ readelf -S <path-to-binary>
  ```

- objdump utility

  ```
  $ objdump -t <path-to-binary>
  ```

Examining Specific Sections

By far the most frequently examined sections are the ones containing the linker symbols. For that reason, a wide variety of tools has been developed to address this specific need. For the same reason, even though it belongs under the broad category of examining the sections, the paragraph describing the symbols extraction has been presented first as a separate topic.

Examining the Dynamic Section

The dynamic section of the binary (the dynamic library in particular) contains plenty of interesting information. Listing the contents of this particular section may be accomplished by relying on one of the following:

- readelf utility

  ```
  $ readelf -d <path-to-binary>
  ```

- objdump utility

  ```
  $ objdump -p <path-to-binary>
  ```

Among the useful pieces of information that may be extracted from dynamic section, here are the ones that are extremely valuable:

- The values of DT_RPATH or DT_RUNPATH fields
- The value of the dynamic library SONAME field
- The list of required dynamic libraries (DT_NEEDED field)

Determining Whether Dynamic Library is PIC or LTR

If the dynamic library is built without the -fPIC compiler flag, its dynamic section features the TEXTREL field, which otherwise would not be present. The following simple script (pic_or_ltr.sh) can help you determine whether the dynamic library was built with -fPIC flag or not:

```
if readelf -d $1 | grep TEXTREL > /dev/null; \
then echo "library is LTR, built without the -fPIC flag"; \
else echo "library was built with -fPIC flag"; fi
```

Examining the Relocation Section

This task may be accomplished by relying on the following:

- readelf utility

  ```
  $ readelf -r <path-to-binary>
  ```

- objdump utility

  ```
  $ objdump -R <path-to-binary>
  ```

Examining the Data Section

This task may be accomplished by relying on the following:

- readelf utility

  ```
  $ readelf -x <section name> <path-to-binary>
  ```

- objdump utility

  ```
  $ objdump -s -j <section name> <path-to-binary>
  ```

List and Examine Segments

This task may be accomplished by relying on the following:

- readelf utility

  ```
  $ readelf --segments <path-to-binary>
  ```

- objdump utility

  ```
  $ objdump -p <path-to-binary>
  ```

Disassembling the Code

In this section, you will examine different approaches to disassembling the code.

Disassembling the Binary File

The best tool for this particular task is the objdump command. In fact, this is probably the only case in which readelf does not provide a parallel solution. In particular, the .text section may be disassembled by running

```
$ objdump -d <path-to-binary>
```

Additionally, you may specify the printout flavor (AT&T vs. Intel).

```
$ objdump -d -M intel <path-to-binary>
```

If you want to see the source code (if available) interspersed with the assembly instructions, you may run the following:

```
$ objdump -d -M intel -S <path-to-binary>
```

Finally, you may want to analyze the code in a given section. Other than the .text section, which is notorious for carrying code, some other sections (.plt, for example) can contain source code.

By default, objdump disassembles all the sections carrying code. To specify the individual section to disassemble, use -j option:

```
$ objdump -d -S -M intel -j .plt <path-to-binary>
```

Disassembling the Running Process

The best way is to rely on the gdb debugger. Please refer to the previous chapter's section devoted to this wonderful tool.

Identifying the Debug Build

It seems that the most reliable way to recognize whether the binary has been built for debug (i.e., with the -g option) is to rely on the readelf tool. In particular, running

```
$ readelf --debug-dump=line <path-to-binary>
```

will provide non-empty output in the case of debug version of the binary file.

Listing Load-time Dependencies

To list the set of shared libraries on which an executable (application and/or shared library) depends on at load time, please take a closer look at the discussion devoted to the ldd command (in which both the ldd method and a safer method based on the objdump) have been mentioned).

In a nutshell, running ldd

```
$ ldd <path-to-binary>
```

will provide the complete list of dependencies.

Alternatively, relying on objdump or readelf to examine the dynamic section of the binaries is a bit safer proposition, which comes at the cost of providing only the first level of dependencies.

```
$ objdump -p /path/to/program | grep NEEDED
$ readelf -d /path/to/program | grep NEEDED
```

Listing the Libraries Known to the Loader

To list all the libraries whose runtime paths are known and available to the loader, you may rely on the ldconfig utility. Running

```
$ ldconfig -p
```

will print the complete list of libraries known to the loader (i.e., currently present in the /etc/ld.so.cache file) together with their respective paths.

Consequently, searching for a particular library in the entire list of libraries available to the loader can be accomplished by running

```
$ ldconfig -p | grep <library-of-interest>
```

Listing Dynamically Linked Libraries

As opposed to the tasks listed so far in this chapter, this particular task has not been mentioned in the context of the binary analysis tools. The reason is simple: the binary file analysis tools are of little use at runtime when the runtime dynamic library loading happens. Tools such as ldd do not cover the dynamic libraries loaded at runtime by the call to dlopen() function.

The following methods will provide the complete list of dynamic libraries loaded. The list includes both the libraries dynamically linked as statically aware as well as the libraries dynamically linked at runtime.

strace Utility

Calling strace <program command line> is a useful method for listing the sequence of system calls among which open() and mmap() are the most interesting for us. This method reveals the complete list of loaded shared libraries. Whenever a shared library is mentioned, typically the few output lines below the mmap() call reveals the loading address.

LD_DEBUG Environment Variable

Given its flexibility and wide array of choices, this option is always on the list of tools for tracking down everything related to the linking/loading process. For this particular problem, the LD_DEBUG=files option may provide plenty of printouts carrying the excessive information about the libraries dynamically loaded at runtime (their names, runtime paths, entry point addresses, etc.).

/proc/<ID>/maps File

Whenever a process runs, the Linux operating system maintains a set of files under the /proc folder, keeping track of the important details related to the process. In particular, for the process whose PID is NNNN, the file at location /proc/<NNNN>/maps contains the list of libraries and their respective load addresses. For example, Figure 13-3 shows what this method reports for the Firefox browser.

```
milan@milan$ ps -ef | grep firefox
milan    15536 14480  8 22:57 pts/0     00:00:07 /usr/lib/firefox/firefox
milan    15596 14480  0 22:58 pts/0     00:00:00 grep --color=auto firefox
milan@milan$ cat /proc/15536/maps
a2c00000-a2d00000 rw-p 00000000 00:00 0
a2e00000-a2f00000 rw-p 00000000 00:00 0
a2ffc000-a2ffd000 ---p 00000000 00:00 0
a2ffd000-a37fd000 rw-p 00000000 00:00 0
a37fd000-a37fe000 ---p 00000000 00:00 0
a37fe000-a3ffe000 rw-p 00000000 00:00 0
a3ffe000-a3fff000 ---p 00000000 00:00 0
a3fff000-a47ff000 rw-p 00000000 00:00 0
a47ff000-a4800000 ---p 00000000 00:00 0
a4800000-a5100000 rw-p 00000000 00:00 0

                 o
                 o
                 o
                 o
                 o

a9964000-a999c000 r-xp 00000000 08:01 7868984   /usr/lib/i386-linux-gnu/libcroco-0.6.so.3.0.1
a999c000-a999d000 ---p 00038000 08:01 7868984   /usr/lib/i386-linux-gnu/libcroco-0.6.so.3.0.1
a999d000-a999f000 r--p 00038000 08:01 7868984   /usr/lib/i386-linux-gnu/libcroco-0.6.so.3.0.1
a999f000-a99a0000 rw-p 0003a000 08:01 7868984   /usr/lib/i386-linux-gnu/libcroco-0.6.so.3.0.1
a99a0000-a99d7000 r-xp 00000000 08:01 7869354   /usr/lib/i386-linux-gnu/librsvg-2.so.2.36.1
a99d7000-a99d8000 r--p 00036000 08:01 7869354   /usr/lib/i386-linux-gnu/librsvg-2.so.2.36.1
a99d8000-a99d9000 rw-p 00037000 08:01 7869354   /usr/lib/i386-linux-gnu/librsvg-2.so.2.36.1
a99d9000-a9a00000 r--p 00000000 08:01 9568812   /usr/share/xul-ext/ubufox/chrome/ubufox.jar
```

Figure 13-3. Examining /proc/<PID>/maps file to examine process memory map

REMARK 1:

A potential small problem may be that certain applications complete quickly, not leaving enough time to examine the process memory map. The simplest and quickest solution in this case would be to start the process through the gdb debugger and put a breakpoint on the main function. While the program execution stays blocked on the breakpoint, you will have unlimited time to examine the process memory map.

REMARK 2:

If you are sure that only one instance of the program is currently being executed, you can eliminate the need to look for the process PID by relying on the pgrep (process grep) command. In the case of Firefox browser, you would type

```
$ cat /proc/`pgrep firefox`/maps
```

lsof Utility

The lsof utility analyses the running process and prints out in the standard output stream the list of all files opened by a process. As stated in its man page (http://linux.die.net/man/8/lsof), an open file may be a regular file, a directory, a block special file, a character special file, an executing text reference, a library, a stream, or a network file (Internet socket, NFS file, or UNIX domain socket).

Among the broad selection of file types it reports being open, it also reports the list of dynamic libraries loaded by the process, regardless of whether the loading was performed statically aware or dynamically (by running dlopen at runtime).

The following snipped illustrates how to get the list of all shared libraries opened by the Firefox browser shown in Figure 13-4:

```
$ lsof -p `pgrep firefox`
```

```
milan@milan:~/Desktop$ ps -ef | grep firefox
milan     3463  2625 10 20:59 pts/3    00:00:01 /usr/lib/firefox/firefox
milan     3506  2625  0 20:59 pts/3    00:00:00 grep --color=auto firefox
milan@milan:~/Desktop$ lsof -p 3463 | grep "\.so"
firefox 3463 milan  mem    REG        8,1    458376 7867630 /usr/lib/firefox/libnssckbi.so
firefox 3463 milan  mem    REG        8,1    394592 7867626 /usr/lib/firefox/libfreebl3.so
firefox 3463 milan  mem    REG        8,1     22080 7344057 /lib/i386-linux-gnu/libnss_dns-2.15.so
firefox 3463 milan  mem    REG        8,1    268144 7867915 /usr/lib/firefox/libsoftokn3.so
firefox 3463 milan  mem    REG        8,1    161096 7867492 /usr/lib/firefox/libnssdbm3.so
firefox 3463 milan  mem    REG        8,1    239248 7868934 /usr/lib/i386-linux-gnu/libcroco-0.6.so.3.0.1
firefox 3463 milan  mem    REG        8,1    227972 7869354 /usr/lib/i386-linux-gnu/librsvg-2.so.2.36.1
                                 o
                                 o
                                 o
                                 o
                                 o
                                 o
firefox 3463 milan  mem    REG        8,1    905712 7869397 /usr/lib/i386-linux-gnu/libstdc++.so.6.0.16
firefox 3463 milan  mem    REG        8,1     30684 7344054 /lib/i386-linux-gnu/librt-2.15.so
firefox 3463 milan  mem    REG        8,1     13940 7344062 /lib/i386-linux-gnu/libdl-2.15.so
firefox 3463 milan  mem    REG        8,1    124663 7344052 /lib/i386-linux-gnu/libpthread-2.15.so
firefox 3463 milan  mem    REG        8,1      5408 7865724 /usr/lib/i386-linux-gnu/libgthread-2.0.so.0.3200.4
firefox 3463 milan  mem    REG        8,1      9624 7867962 /usr/lib/firefox/libmozalloc.so
firefox 3463 milan  mem    REG        8,1     13604 7867057 /usr/lib/firefox/libplds4.so
firefox 3463 milan  mem    REG        8,1     17700 7867631 /usr/lib/firefox/libplc4.so
firefox 3463 milan  mem    REG        8,1    134344 7344053 /lib/i386-linux-gnu/ld-2.15.so
milan@milan:~/Desktop$
```

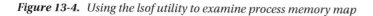

Figure 13-4. Using the lsof utility to examine process memory map

Note that lsof provides the command line option of running the process examination periodically. By specifying the examination period you may catch the moments in which the runtime dynamic loading and unloading happens.

When running lsof with the -r option, the periodic process examination goes on in an endless loop, requiring the user to press Ctrl-C to terminate. Running lsof with the +r option has the effect of lsof terminating when no more open files are detected.

Programmatic Way

It is also possible to write code so that it prints out the libraries being loaded by the process. When application code incorporates calls to the dl_iterate_phdr() function, its printouts at runtime may help you determine the complete list of shared libraries that it loads as well as the extra data associated with each library (such as the loaded library start address).

To illustrate the concept, demo code comprised of a driver app and two simple dynamic libraries has been created. The app's source file is shown in in the following example. One of the dynamic libraries is dynamically linked statically aware, whereas the other library is dynamically loaded by invoking the dlopen() function.

```
#define _GNU_SOURCE
#include <link.h>
#include <stdio.h>
#include <dlfcn.h>
#include "sharedLib1Functions.h"
#include "sharedLib2Functions.h"

static const char* segment_type_to_string(uint32_t type)
{
    switch(type)
    {
    case PT_NULL:        // 0
        return "Unused";
        break;
    case PT_LOAD:        // 1
        return "Loadable Program Segment";
        break;
    case PT_DYNAMIC:     //2
        return "Dynamic linking information";
        break;
    case PT_INTERP:      // 3
        return "Program interpreter";
        break;
    case PT_NOTE:        // 4
        return "Auxiliary information";
        break;
    case PT_SHLIB:       // 5
        return "Reserved";
        break;
    case PT_PHDR:        // 6
        return "Entry for header table itself";
        break;
    case PT_TLS:         // 7
        return "Thread-local storage segment";
        break;
//  case PT_NUM:         // 8                   /* Number of defined types */
    case PT_LOOS:        // 0x60000000
        return "Start of OS-specific";
        break;
```

```
        case PT_GNU_EH_FRAME: // 0x6474e550
            return "GCC .eh_frame_hdr segment";
            break;
        case PT_GNU_STACK:      // 0x6474e551
            return "Indicates stack executability";
            break;
        case PT_GNU_RELRO:      // 0x6474e552
            return "Read-only after relocation";
            break;
//  case PT_LOSUNW:         // 0x6ffffffa
        case PT_SUNWBSS:        // 0x6ffffffa
            return "Sun Specific segment";
            break;
        case PT_SUNWSTACK:      // 0x6ffffffb
            return "Sun Stack segment";
            break;
//  case PT_HISUNW:         // 0x6fffffff
//          case PT_HIOS:           // 0x6fffffff        /* End of OS-specific */
//          case PT_LOPROC:         // 0x70000000        /* Start of processor-specific */
//          case PT_HIPROC:         // 0x7fffffff        /* End of processor-specific */
        default:
            return "???";
    }
}

static const char* flags_to_string(uint32_t flags)
{
    switch(flags)
    {
    case 1:
        return "--x";
        break;
    case 2:
        return "-w-";
        break;
    case 3:
        return "-wx";
        break;
    case 4:
        return "r--";
        break;
    case 5:
        return "r-x";
        break;
    case 6:
        return "rw-";
        break;
    case 7:
        return "rwx";
        break;
```

```
        default:
            return "???";
            break;
    }
}

static int header_handler(struct dl_phdr_info* info, size_t size, void* data)
{
    int j;
    printf("name=%s (%d segments) address=%p\n",
            info->dlpi_name, info->dlpi_phnum, (void*)info->dlpi_addr);
    for (j = 0; j < info->dlpi_phnum; j++) {
        printf("\t\t header %2d: address=%10p\n", j,
            (void*) (info->dlpi_addr + info->dlpi_phdr[j].p_vaddr));
        printf("\t\t\t type=0x%X (%s),\n\t\t\t flags=0x%X (%s)\n",
                info->dlpi_phdr[j].p_type,
                segment_type_to_string(info->dlpi_phdr[j].p_type),
                info->dlpi_phdr[j].p_flags,
                flags_to_string(info->dlpi_phdr[j].p_flags));
    }
    printf("\n");
    return 0;
}

int main(int argc, char* argv[])
{
    // function from statically aware loaded library
    sharedLib1Function(argc);

    // function from run-time dynamically loaded library
    void* pLibHandle = dlopen("libdemo2.so", RTLD_GLOBAL | RTLD_NOW);
    if(NULL == pLibHandle)
    {
        printf("Failed loading libdemo2.so, error = %s\n", dlerror());
        return -1;
    }
    PFUNC pFunc = (PFUNC)dlsym(pLibHandle, "sharedLib2Function");
    if(NULL == pFunc)
    {
        printf("Failed identifying the symbol \"sharedLib2Function\"\n");
        dlclose(pLibHandle);
        pLibHandle = NULL;
        return -1;
    }
    pFunc(argc);
    if(2 == argc)
        getchar();
    if(3 == argc)
        dl_iterate_phdr(header_handler, NULL);
    return 0;
}
```

The central place in this code example belongs to the call to the dl_iterate_phdr() function. This function essentially extracts the relevant process mapping information at runtime and passes it to the caller. The caller is responsible for providing the custom implementation of the callback function (header_handler() in this example). Figure 13-5 shows what the produced screen printout may look like.

```
milan@milan$ ./driverApp 1 2 | grep -A 20 libdemo
name=../sharedLib1/libdemo1.so (7 segments) address=0xb77ad000
                    header   0: address=0xb77ad000
                             type=0x1 (Loadable Program Segment),
                             flags=0x5 (r-x)
                    header   1: address=0xb77aef0c
                             type=0x1 (Loadable Program Segment),
                             flags=0x6 (rw-)
                    header   2: address=0xb77aef20
                             type=0x2 (Dynamic linking information),
                             flags=0x6 (rw-)
                    header   3: address=0xb77ad114
                             type=0x4 (Auxiliary information),
                             flags=0x4 (r--)
                    header   4: address=0xb77ad4f8
                             type=0x6474E550 (GCC .eh_frame_hdr segment),
                             flags=0x4 (r--)
                    header   5: address=0xb77ad000
                             type=0x6474E551 (Indicates stack executability),
                             flags=0x6 (rw-)
                    header   6: address=0xb77aef0c
                             type=0x6474E552 (Read-only after relocation),
--
name=../sharedLib2/libdemo2.so (7 segments) address=0xb77c3000
                    header   0: address=0xb77c3000
                             type=0x1 (Loadable Program Segment),
                             flags=0x5 (r-x)
                    header   1: address=0xb77c4f0c
                             type=0x1 (Loadable Program Segment),
                             flags=0x6 (rw-)
                    header   2: address=0xb77c4f20
                             type=0x2 (Dynamic linking information),
                             flags=0x6 (rw-)
                    header   3: address=0xb77c3114
                             type=0x4 (Auxiliary information),
                             flags=0x4 (r--)
                    header   4: address=0xb77c34f8
                             type=0x6474E550 (GCC .eh_frame_hdr segment),
                             flags=0x4 (r--)
                    header   5: address=0xb77c3000
                             type=0x6474E551 (Indicates stack executability),
                             flags=0x6 (rw-)
                    header   6: address=0xb77c4f0c
                             type=0x6474E552 (Read-only after relocation),
milan@milan$
```

Figure 13-5. *The programmatic way (relying on dl_iterate_phdr() call) of examining the dynamic library loading locations in the process memory map*

Creating and Maintaining the Static Library

The majority of the tasks related to dealing specifically with the static libraries can be completed by using the Linux ar archiver. Completing tasks such as disassembling the static library code or inspecting its symbols does not differ from how it is performed on the applications or dynamic libraries.

CHAPTER 14

■■■

Windows Toolbox

The purpose of this chapter is to introduce the reader to the set of tools (utility programs as well as other methods) for analyzing the contents of the Windows binary files. Even though the Linux objdump utility has some capabilities of analyzing the PE/COFF format, the strict focus in this chapter will be on indigenous Windows tools, which are more likely to be up to par with any changes to the PE/COFF format which may happen along the way.

Library Manager (lib.exe)

The Windows 32-bit Library Manager lib.exe comes as a standard part of the Visual Studio development tools (Figure 14-1).

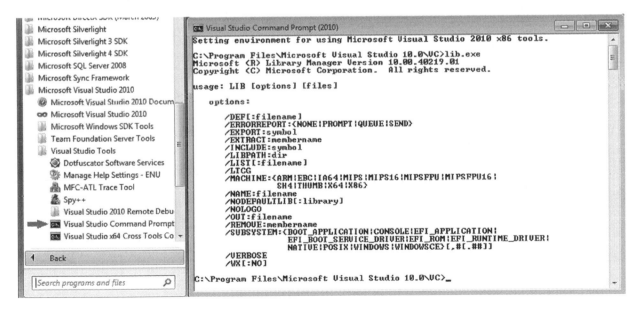

Figure 14-1. *Using the lib.exe utility*

This utility program not only handles the static libraries much in the same way as its Linux counterpart (archiver ar), but also plays a role in the domain of dynamic libraries as the tool that can create import libraries (the collection of DLL symbols, file extension .lib) as well as the export files (capable of resolving the circular dependencies, file extension .exp). Detailed documentation about lib.exe can be found at the MSDN site (http://msdn.microsoft.com/en-us/library/7ykb2k5f.aspx).

lib.exe as a Static Library Tool

In this section I will illustrate the typical roles in which the lib.exe tool may be really useful.

lib.exe as a Default Archiver Tool

When Visual Studio is used to create a C/C++ static library project, lib.exe is set as default archiver/librarian tool, and the project settings' Librarian tab is used to specify the command line options for it (Figure 14-2).

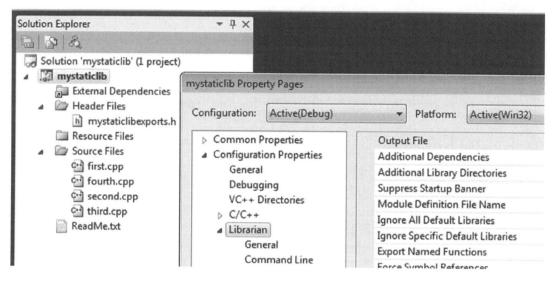

Figure 14-2. *Using lib.exe as default archiver*

By default, building the static library project invokes lib.exe after the compilation stage, which happens without any action required by the developer. However, this is not necessarily where the use of lib.exe must end. It is possible to run lib.exe from the Visual Studio command prompt much in the same way as the Linux ar archiver is used to perform the same kinds of tasks.

lib.exe as a Command Line Utility

In order to illustrate the use of lib.exe you will create a Windows static library fully matching the functionality of the Linux static library used to demonstrate the use of ar in Chapter 10. The demo project is comprised of four source files (first.c, second.c, third.c, and fourth.c) and one export header file which can be used by the client binaries. These files are shown in the following five examples.

file: first.c
```
#include "mystaticlibexports.h"

int first_function(int x)
{
        return (x+1);
}
```

file: second.c
```
#include "mystaticlibexports.h"

int fourth_function(int x)
{
        return (x+4);
}
```

file: third.c
```
#include "mystaticlibexports.h"

int second_function(int x)
{
        return (x+2);
}
```

file: fourth.c
```
#include "mystaticlibexports.h"

int third_function(int x)
{
        return (x+3);
}
```
file: mystaticlibexports.h
```
#pragma once
int first_function(int x);
int second_function(int x);
int third_function(int x);
int fourth_function(int x);
```

Creating a Static Library

Let's assume that you compiled all four source files and that you have available four object files (first.obj, second.obj, third.obj, and fourth.obj). Passing the desired library name to lib.exe (after the /OUT flag) followed by the list of participating object files will have the effect of creating the static library, as shown in Figure 14-3.

```
c:\Users\milan\mystaticlib\mystaticlib\Debug>dir *.obj
 Volume in drive C has no label.
 Volume Serial Number is F4F7-CFD4

 Directory of c:\Users\milan\mystaticlib\mystaticlib\Debug

12/25/2013  06:48 PM              2,626 first.obj
12/25/2013  06:48 PM              2,635 fourth.obj
12/25/2013  06:48 PM              2,635 second.obj
12/25/2013  06:48 PM              2,626 third.obj
               4 File(s)         10,522 bytes
               0 Dir(s)  132,602,314,752 bytes free

c:\Users\milan\mystaticlib\mystaticlib\Debug>lib.exe /OUT:mystaticlib.lib
/NOLOGO first.obj second.obj third.obj fourth.obj

c:\Users\milan\mystaticlib\mystaticlib\Debug>dir *.lib
 Volume in drive C has no label.
 Volume Serial Number is F4F7-CFD4

 Directory of c:\Users\milan\mystaticlib\mystaticlib\Debug

12/25/2013  06:50 PM             11,140 mystaticlib.lib
               1 File(s)         11,140 bytes
               0 Dir(s)  132,602,302,464 bytes free

c:\Users\milan\mystaticlib\mystaticlib\Debug>
```

Figure 14-3. *Using lib.exe to combine object files into a static library*

In order to completely mimic the default settings supplied by Visual Studio when creating the static library project, I've added the /NOLOGO argument.

Listing the Static Library Contents

When the /LIST flag is passed to lib.exe, it prints out the list of object files currently contained by the static library, as shown in Figure 14-4.

```
c:\Users\milan\mystaticlib\mystaticlib\Debug>lib.exe /LIST mystaticlib.lib
Microsoft (R) Library Manager Version 10.00.40219.01
Copyright (C) Microsoft Corporation.  All rights reserved.

first.obj
second.obj
third.obj
fourth.obj

c:\Users\milan\mystaticlib\mystaticlib\Debug>
```

Figure 14-4. *Using lib.exe to list the object files of static library*

Removing Individual Object Files from the Static Library

The individual object files can be removed from the static library by passing the /REMOVE flag to lib.exe (Figure 14-5).

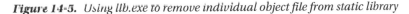

```
c:\Users\milan\mystaticlib\mystaticlib\Debug>lib.exe /REMOVE:first.obj mystaticlib.lib

Microsoft (R) Library Manager Version 10.00.40219.01
Copyright (C) Microsoft Corporation.  All rights reserved.


c:\Users\milan\mystaticlib\mystaticlib\Debug>lib.exe /LIST mystaticlib.lib
Microsoft (R) Library Manager Version 10.00.40219.01
Copyright (C) Microsoft Corporation.  All rights reserved.

fourth.obj
third.obj
second.obj

c:\Users\milan\mystaticlib\mystaticlib\Debug>
```

Figure 14-5. Using lib.exe to remove individual object file from static library

Inserting the Object File into the Static Library

The new object file may be added to the existing static library by passing the library filename followed by the list of object files to be added. This syntax is very similar to the scenario of creating the static library, except that the /OUT flag may be omitted (Figure 14-6).

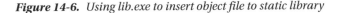

```
c:\Users\milan\mystaticlib\mystaticlib\Debug>lib.exe /LIST mystaticlib.lib
Microsoft (R) Library Manager Version 10.00.40219.01
Copyright (C) Microsoft Corporation.  All rights reserved.

fourth.obj
third.obj
second.obj

c:\Users\milan\mystaticlib\mystaticlib\Debug>lib.exe mystaticlib.lib first.obj
Microsoft (R) Library Manager Version 10.00.40219.01
Copyright (C) Microsoft Corporation.  All rights reserved.


c:\Users\milan\mystaticlib\mystaticlib\Debug>lib.exe /LIST mystaticlib.lib
Microsoft (R) Library Manager Version 10.00.40219.01
Copyright (C) Microsoft Corporation.  All rights reserved.

first.obj
second.obj
third.obj
fourth.obj

c:\Users\milan\mystaticlib\mystaticlib\Debug>
```

Figure 14-6. Using lib.exe to insert object file to static library

Extracting the Individual Object File from the Static Library

Finally, the individual object files may be extracted from the static library. In order to demonstrate it, I first purposefully deleted the original object file (first.obj) whose extraction from the static library is planned to happen (Figure 14-7).

```
c:\Users\milan\mystaticlib\mystaticlib\Debug>dir *.obj
 Volume in drive C has no label.
 Volume Serial Number is F4F7-CFD4

 Directory of c:\Users\milan\mystaticlib\mystaticlib\Debug

12/25/2013  06:59 PM             2,626 first.obj
12/25/2013  06:58 PM             2,635 fourth.obj
12/25/2013  06:58 PM             2,635 second.obj
12/25/2013  06:58 PM             2,626 third.obj
               4 File(s)         10,522 bytes
               0 Dir(s) 132,601,217,024 bytes free

c:\Users\milan\mystaticlib\mystaticlib\Debug>del first*.obj

c:\Users\milan\mystaticlib\mystaticlib\Debug>dir *.obj
 Volume in drive C has no label.
 Volume Serial Number is F4F7-CFD4

 Directory of c:\Users\milan\mystaticlib\mystaticlib\Debug

12/25/2013  06:58 PM             2,635 fourth.obj
12/25/2013  06:58 PM             2,635 second.obj
12/25/2013  06:58 PM             2,626 third.obj
               3 File(s)          7,896 bytes
               0 Dir(s) 132,601,221,120 bytes free

c:\Users\milan\mystaticlib\mystaticlib\Debug>lib.exe /LIST mystaticlib.lib
Microsoft (R) Library Manager Version 10.00.40219.01
Copyright (C) Microsoft Corporation.  All rights reserved.

first.obj
second.obj
third.obj
fourth.obj

c:\Users\milan\mystaticlib\mystaticlib\Debug>lib.exe /EXTRACT:first.obj
mystaticlib.lib
Microsoft (R) Library Manager Version 10.00.40219.01
Copyright (C) Microsoft Corporation.  All rights reserved.

c:\Users\milan\mystaticlib\mystaticlib\Debug>dir *.obj
 Volume in drive C has no label.
 Volume Serial Number is F4F7-CFD4

 Directory of c:\Users\milan\mystaticlib\mystaticlib\Debug

12/25/2013  06:59 PM             2,626 first.obj
12/25/2013  06:58 PM             2,635 fourth.obj
12/25/2013  06:58 PM             2,635 second.obj
12/25/2013  06:58 PM             2,626 third.obj
               4 File(s)         10,522 bytes
               0 Dir(s) 132,601,217,024 bytes free

c:\Users\milan\mystaticlib\mystaticlib\Debug>
```

Figure 14-7. Using lib.exe to extract an individual object file from the static library

lib.exe in the Realm of Dynamic Libraries (Import Library Tool)

lib.exe is also used to create the DLL import library (.lib) file and export file (.exp) based on the available export definition file (.def). When working strictly within the Visual Studio environment, this task is typically automatically assigned to lib.exe. A far more interesting scenario happens when the DLL is created by a third-party compiler which does not create the corresponding import library and export file. In such cases, lib.exe must be run from the command line (i.e., the Visual Studio command prompt).

The following example illustrates how lib.exe can be used to create the missing import libraries after the cross compiling session in which MinGW compiler run on Linux produced the Windows binaries, but did not supply the needed import libraries (Figure 14-8).

```
X:\MilanFFMpegWin32Build>dir *.def
 Volume in drive X is VBOX_VBoxShared
 Volume Serial Number is 9AE7-0879

 Directory of X:\WinFFMpegBuiltOnLinux

02/14/2013  11:51 AM             7,012 avcodec-53.def
02/14/2013  11:51 AM               115 avdevice-53.def
02/14/2013  11:51 AM             5,107 avfilter-2.def
02/14/2013  11:51 AM             5,119 avformat-53.def
02/14/2013  11:51 AM             4,762 avutil-51.def
02/14/2013  11:51 AM               232 postproc-51.def
02/14/2013  11:51 AM               155 swresample-0.def
02/14/2013  11:51 AM             7,084 swscale-2.def
               8 File(s)         29,586 bytes
               0 Dir(s)  465,080,082,432 bytes free

X:\MilanFFMpegWin32Build>lib /machine:X86 /def:avcodec-53.def /out:avcodec.lib

Microsoft (R) Library Manager Version 10.00.40219.01
Copyright (C) Microsoft Corporation.  All rights reserved.

   Creating library avcodec.lib and object avcodec.exp

X:\MilanFFMpegWin32Build>lib /machine:X86 /def:avdevice-53.def /out:avdevice.lib

Microsoft (R) Library Manager Version 10.00.40219.01
Copyright (C) Microsoft Corporation.  All rights reserved.

   Creating library avdevice.lib and object avdevice.exp

X:\MilanFFMpegWin32Build>lib /machine:X86 /def:avfilter-2.def /out:avfilter.lib

Microsoft (R) Library Manager Version 10.00.40219.01
Copyright (C) Microsoft Corporation.  All rights reserved.

   Creating library avfilter.lib and object avfilter.exp

X:\MilanFFMpegWin32Build>lib /machine:X86 /def:avformat-53.def /out:avformat.lib

Microsoft (R) Library Manager Version 10.00.40219.01
Copyright (C) Microsoft Corporation.  All rights reserved.

   Creating library avformat.lib and object avformat.exp

X:\MilanFFMpegWin32Build>lib /machine:X86 /def:avutil-51.def /out:avutil.lib
Microsoft (R) Library Manager Version 10.00.40219.01
Copyright (C) Microsoft Corporation.  All rights reserved.

   Creating library avutil.lib and object avutil.exp

X:\MilanFFMpegWin32Build>lib /machine:X86 /def:postproc-51.def /out:postproc.lib

Microsoft (R) Library Manager Version 10.00.40219.01
Copyright (C) Microsoft Corporation.  All rights reserved.

   Creating library postproc.lib and object postproc.exp

X:\MilanFFMpegWin32Build>lib /machine:X86 /def:swresample-0.def /out:swresample.
lib
Microsoft (R) Library Manager Version 10.00.40219.01
Copyright (C) Microsoft Corporation.  All rights reserved.

   Creating library swresample.lib and object swresample.exp

X:\MilanFFMpegWin32Build>lib /machine:X86 /def:swscale-2.def /out:swscale.lib
Microsoft (R) Library Manager Version 10.00.40219.01
Copyright (C) Microsoft Corporation.  All rights reserved.

   Creating library swscale.lib and object swscale.exp

X:\MilanFFMpegWin32Build>
```

Figure 14-8. *Using lib.exe to create an import library based on DLL and its definition (.DEF)_file*

dumpbin Utility

The Visual Studio dumpbin utility (http://support.microsoft.com/kb/177429) is for the most part the Windows equivalent of Linux objdump utility, as it performs the examination and the analysis of the important details of executable, such as exported symbols, sections, disassembling the code (.text) sections, list of object files in static library, etc.

This tool is also a standard part of the Visual Studio package. Similar to the previously described lib tool, it is standardly run from the Visual Studio Command Prompt (Figure 14-9).

Figure 14-9. Using the dumpbin utility

The typical tasks described in the following sections can be completed by running dumpbin.

Identifying the Binary File Type

When run without the extra flags, dumpbin reports the binary file type (Figure 14-10).

```
c:\Users\milan\DLLVersioningDemo\VersionedDLL\Debug>dumpbin dllmain.obj
Microsoft (R) COFF/PE Dumper Version 10.00.40219.01
Copyright (C) Microsoft Corporation.  All rights reserved.


Dump of file dllmain.obj

File Type: COFF OBJECT

  Summary

          4 .bss
       1F50 .debug$S
         64 .debug$T
         41 .drectve
          4 .rtc$IMZ
          4 .rtc$IMZ
         5D .text

c:\Users\milan\DLLVersioningDemo\VersionedDLL\Debug>cd ..\..\Debug

c:\Users\milan\DLLVersioningDemo\Debug>dumpbin VersionedDLL.dll
Microsoft (R) COFF/PE Dumper Version 10.00.40219.01
Copyright (C) Microsoft Corporation.  All rights reserved.


Dump of file VersionedDLL.dll

File Type: DLL

  Summary

       1000 .data
       1000 .idata
       2000 .rdata
       1000 .reloc
       1000 .rsrc
       4000 .text
      10000 .textbss

c:\Users\milan\DLLVersioningDemo\Debug>dumpbin VersionedDLLClientApp.exe
Microsoft (R) COFF/PE Dumper Version 10.00.40219.01
Copyright (C) Microsoft Corporation.  All rights reserved.


Dump of file VersionedDLLClientApp.exe

File Type: EXECUTABLE IMAGE

  Summary

       1000 .data
       1000 .idata
       2000 .rdata
       1000 .reloc
       1000 .rsrc
       4000 .text
      10000 .textbss

c:\Users\milan\DLLVersioningDemo\Debug>
```

Figure 14-10. *Using the dumpbin utility to identify binary file types*

Listing the DLL Exported Symbols

Running dumpbin /EXPORTS <dll path> provides the list of exported symbols (Figure 14-11).

```
c:\Users\milan\DLLVersioningDemo\Debug>dumpbin /EXPORTS VersionedDLL.dll
Microsoft (R) COFF/PE Dumper Version 10.00.40219.01
Copyright (C) Microsoft Corporation.  All rights reserved.


Dump of file VersionedDLL.dll

File Type: DLL

  Section contains the following exports for VERSIONEDDLL.dll

    00000000 characteristics
    52B625A0 time date stamp Sat Dec 21 15:34:56 2013
        0.00 version
           1 ordinal base
           1 number of functions
           1 number of names

    ordinal hint RVA       name

          1    0 00011087 DllGetVersion = @ILT+130(?DllGetVersion@@YGJPAU_DLLVER
SIONINFO@@@Z)

  Summary

        1000 .data
        1000 .idata
        2000 .rdata
        1000 .reloc
        1000 .rsrc
        4000 .text
       10000 .textbss

c:\Users\milan\DLLVersioningDemo\Debug>
```

Figure 14-11. Using dumpbin utility to list exported symbols of DLL file

Listing and Examining the Sections

Running dumpbin /HEADERS <binary file path> prints out the complete list of sections present in the file (Figure 14-12).

```
c:\Users\milan\DLLVersioningDemo\Debug>dumpbin /HEADERS VersionedDLL.dll
Microsoft (R) COFF/PE Dumper Version 10.00.40219.01
Copyright (C) Microsoft Corporation.  All rights reserved.


Dump of file VersionedDLL.dll

PE signature found

File Type: DLL

FILE HEADER VALUES
             14C machine (x86)
               7 number of sections
        52B697A6 time date stamp Sat Dec 21 23:41:26 2013
               0 file pointer to symbol table
               0 number of symbols
              E0 size of optional header
            2102 characteristics
                   Executable
                   32 bit word machine
                   DLL

OPTIONAL HEADER VALUES
             10B magic # (PE32)
                         o
                         o
                         o
SECTION HEADER #1
.textbss name
   10000 virtual size
    1000 virtual address (10001000 to 10010FFF)
       0 size of raw data
       0 file pointer to raw data
       0 file pointer to relocation table
       0 file pointer to line numbers
       0 number of relocations
       0 number of line numbers
E00000A0 flags
         Code
         Uninitialized Data
         Execute Read Write

SECTION HEADER #2
   .text name
    3CA3 virtual size
   11000 virtual address (10011000 to 10014CA2)
    3E00 size of raw data
     400 file pointer to raw data (00000400 to 000041FF)
       0 file pointer to relocation table
       0 file pointer to line numbers
       0 number of relocations
       0 number of line numbers
60000020 flags
         Code
         Execute Read

                         o
                         o
                         o
```

Figure 14-12. *Using dumpbin to list the sections*

Once the section names are listed, the individual section info can be obtained by running dumpbin /SECTION:<section name> <binary file path> (Figure 14-13).

```
c:\Users\milan\DLLVersioningDemo\Debug>dumpbin /SECTION:.text VersionedDLL.dll
Microsoft (R) COFF/PE Dumper Version 10.00.40219.01
Copyright (C) Microsoft Corporation.  All rights reserved.


Dump of file VersionedDLL.dll

File Type: DLL

SECTION HEADER #2
   .text name
    3CA3 virtual size
   11000 virtual address (10011000 to 10014CA2)
    3E00 size of raw data
     400 file pointer to raw data (00000400 to 000041FF)
       0 file pointer to relocation table
       0 file pointer to line numbers
       0 number of relocations
       0 number of line numbers
60000020 flags
         Code
         Execute Read

  Summary

        4000 .text

c:\Users\milan\DLLVersioningDemo\Debug>dumpbin /SECTION:.data VersionedDLL.dll
Microsoft (R) COFF/PE Dumper Version 10.00.40219.01
Copyright (C) Microsoft Corporation.  All rights reserved.


Dump of file VersionedDLL.dll

File Type: DLL

SECTION HEADER #4
   .data name
     7C0 virtual size
   17000 virtual address (10017000 to 100177BF)
     200 size of raw data
    5E00 file pointer to raw data (00005E00 to 00005FFF)
       0 file pointer to relocation table
       0 file pointer to line numbers
       0 number of relocations
       0 number of line numbers
C0000040 flags
         Initialized Data
         Read Write

  Summary

        1000 .data
c:\Users\milan\DLLVersioningDemo\Debug>
```

Figure 14-13. Using dumpbin to get detailed insight into a specific section

Disassembling the Code

Running dumpbin /DISASM <binary file path> provides the disassembled listing of the complete binary file (Figure 14-14).

```
c:\Users\milan\DLLVersioningDemo\Debug>dumpbin /DISASM VersionedDLL.dll
Microsoft (R) COFF/PE Dumper Version 10.00.40219.01
Copyright (C) Microsoft Corporation.  All rights reserved.


Dump of file VersionedDLL.dll

File Type: DLL

  10011000: CC                      int          3
  10011001: CC                      int          3
  10011002: CC                      int          3
  10011003: CC                      int          3
  10011004: CC                      int          3
@ILT+0(_wcstok_s):
  10011005: E9 04 0A 00 00          jmp          _wcstok_s
@ILT+5(__wtoi):
  1001100A: E9 F9 09 00 00          jmp          __wtoi
@ILT+10(__RTC_GetErrDesc):
  1001100F: E9 2C 13 00 00          jmp          __RTC_GetErrDesc
@ILT+15(__malloc_dbg):
  10011014: E9 E7 1E 00 00          jmp          __malloc_dbg
@ILT+20(@__security_check_cookie@4):
  10011019: E9 D2 2A 00 00          jmp          @__security_check_cookie@4
@ILT+25(_IsDebuggerPresent@0):
  1001101E: E9 99 2C 00 00          jmp          _IsDebuggerPresent@0
@ILT+30(_GetUserDefaultLangID@0):
  10011023: E9 66 09 00 00          jmp          _GetUserDefaultLangID@0
@ILT+35(__RTC_Terminate):
  10011028: E9 A3 1E 00 00          jmp          __RTC_Terminate
@ILT+40(_WideCharToMultiByte@32):
  1001102D: E9 84 2C 00 00          jmp          _WideCharToMultiByte@32
@ILT+45(_DllMain@12):
  10011032: E9 79 03 00 00          jmp          _DllMain@12
                          o
                          o
                          o
  100116B4: E8 91 FA FF FF          call         @ILT+325(__RTC_CheckEsp)
  100116B9: 89 45 A4                mov          dword ptr [ebp-5Ch],eax
  100116BC: 8B F4                   mov          esi,esp
  100116BE: 8B 45 A4                mov          eax,dword ptr [ebp-5Ch]
  100116C1: 50                      push         eax
  100116C2: FF 15 38 83 01 10       call         dword ptr [__imp___wtoi]
  100116C8: 83 C4 04                add          esp,4
  100116CB: 3B F4                   cmp          esi,esp
  100116CD: E8 78 FA FF FF          call         @ILT+325(__RTC_CheckEsp)
  100116D2: 8B 4D 08                mov          ecx,dword ptr [ebp+8]
  100116D5: 89 41 10                mov          dword ptr [ecx+10h],eax
  100116D8: 8B 45 08                mov          eax,dword ptr [ebp+8]
  100116DB: C7 00 14 00 00 00       mov          dword ptr [eax],14h
  100116E1: 8B F4                   mov          esi,esp
  100116E3: 8B 45 BC                mov          eax,dword ptr [ebp-44h]
  100116E6: 50                      push         eax
  100116E7: FF 15 28 82 01 10       call         dword ptr [__imp__FreeResource@4]
  100116ED: 3B F4                   cmp          esi,esp
                          o
                          o
                          o
```

Figure 14-14. Using dumpbin to disassemble the code

Identifying the Debug Build

The dumpbin utility is used to identify the debug version of a binary file. The indicators of a debug build vary depending on the actual binary file type.

Object Files

Running dumpbin /SYMBOLS <binary file path> on the object files (*.obj) will report the object file built for debugging as file of type COFF OBJECT (Figure 14-15).

```
c:\Users\milan\DLLVersioningDemo\VersionedDLL\Debug>dumpbin /SYMBOLS dllmain.obj

Microsoft (R) COFF/PE Dumper Version 10.00.40219.01
Copyright (C) Microsoft Corporation.  All rights reserved.


Dump of file dllmain.obj

File Type: COFF OBJECT

COFF SYMBOL TABLE
000 00AB9D1B ABS    notype         Static       | @comp.id
001 00000001 ABS    notype         Static       | @feat.00
002 00000000 SECT1  notype         Static       | .drectve
    Section length    41, #relocs    0, #linenums    0, checksum        0
    Relocation CRC 00000000
005 00000000 SECT2  notype         Static       | .debug$S
    Section length 1E24, #relocs    2, #linenums    0, checksum        0
    Relocation CRC 8E00A6D9
008 00000000 SECT3  notype         Static       | .bss
    Section length     4, #relocs    0, #linenums    0, checksum        0
    Relocation CRC 00000000
00B 00000000 SECT3  notype         External     | ?g_hModule@@3PAUHINSTANCE__@@A (
struct HINSTANCE__ * g_hModule)
00C 00000000 SECT4  notype         Static       | .text
    Section length    5D, #relocs    2, #linenums    0, checksum A5F7AE6C, select
ion    1 (pick no duplicates)
    Relocation CRC D94D98E5
00F 00000000 SECT5  notype         Static       | .debug$S
    Section length   12C, #relocs    5, #linenums    0, checksum        0, select
ion    5 (pick associative Section 0x4)
    Relocation CRC 638E05AE
012 00000000 SECT4  notype ()      External     | _DllMain@12
013 00000000 SECT6  notype         Static       | .rtc$TMZ
    Section length     4, #relocs    1, #linenums    0, checksum        0, select
ion    5 (pick associative Section 0x4)
    Relocation CRC 4C2E11CC
016 00000000 SECT6  notype         Static       | __RTC_Shutdown.rtc$TMZ
017 00000000 UNDEF  notype ()      External     | __RTC_Shutdown
018 00000000 SECT7  notype         Static       | .rtc$IMZ
    Section length     4, #relocs    1, #linenums    0, checksum        0, select
ion    5 (pick associative Section 0x4)
    Relocation CRC 5D907A9E
01B 00000000 SECT7  notype         Static       | __RTC_InitBase.rtc$IMZ
01C 00000000 UNDEF  notype ()      External     | __RTC_InitBase
01D 00000000 SECT8  notype         Static       | .debug$T
    Section length    64, #relocs    0, #linenums    0, checksum        0
    Relocation CRC 00000000

String Table Size = 0x7B bytes

  Summary

          4 .bss
       1F50 .debug$S
         64 .debug$T
         41 .drectve
          4 .rtc$IMZ
          4 .rtc$TMZ
         5D .text

c:\Users\milan\DLLVersioningDemo\VersionedDLL\Debug>
```

Figure 14-15. *Using dumpbin to detect the debug version of the object file*

The release version of the same file will be reported as file type ANONYMOUS OBJECT (Figure 14-16).

```
c:\Users\milan\DLLVersioningDemo\VersionedDLL\Release>dumpbin /SYMBOLS dllmain.o
bj
Microsoft (R) COFF/PE Dumper Version 10.00.40219.01
Copyright (C) Microsoft Corporation.  All rights reserved.


Dump of file dllmain.obj

File Type: ANONYMOUS OBJECT

c:\Users\milan\DLLVersioningDemo\VersionedDLL\Release>
```

Figure 14-16. *Indication of the release built of the object file*

DLLs and Executables

The certain indicator that a DLL or executable file was built for debugging is the presence of .idata section in the output of running the dumpbin /HEADERS option. The purpose of this section is to support the "edit and continue" feature that is available in debug mode only. More specifically, to enable this option the /INCREMENTAL linker flag is required and typically set for Debug and disabled for Release configuration (Figure 14-17).

```
c:\Users\milan\DLLVersioningDemo\Debug>dumpbin /HEADERS VersionedDLL.dll
Microsoft (R) COFF/PE Dumper Version 10.00.40219.01
Copyright (C) Microsoft Corporation.  All rights reserved.


Dump of file VersionedDLL.dll

PE signature found

File Type: DLL

FILE HEADER VALUES
             14C machine (x86)
               7 number of sections
        52B697A6 time date stamp Sat Dec 21 23:41:26 2013
               0 file pointer to symbol table
               0 number of symbols
              E0 size of optional header
            2102 characteristics
                   Executable
                   32 bit word machine
                   DLL

                     o
                     o
                     o

SECTION HEADER #5
   .idata name
      961 virtual size
    18000 virtual address (10018000 to 10018960)
      A00 size of raw data
     6000 file pointer to raw data (00006000 to 000069FF)
        0 file pointer to relocation table
        0 file pointer to line numbers
        0 number of relocations
        0 number of line numbers
 C0000040 flags
             Initialized Data
             Read Write

                     o
                     o
                     o
```

Figure 14-17. Using dumpbin to detect the debug version of DLL

Listing the Load Time Dependencies

The complete list of the dependency libraries and the symbols imported from them may be obtained by running dumpbin /IMPORTS <binary file path> (Figure 14-18).

```
c:\Users\milan\DLLVersioningDemo\Debug>dumpbin /IMPORTS VersionedDLL.dll
Microsoft (R) COFF/PE Dumper Version 10.00.40219.01
Copyright (C) Microsoft Corporation.  All rights reserved.


Dump of file VersionedDLL.dll

File Type: DLL

  Section contains the following imports:

    VERSION.dll
              100183AC Import Address Table
              100181F0 Import Name Table
                     0 time date stamp
                     0 Index of first forwarder reference

                   E VerQueryValueW

    KERNEL32.dll
              10018220 Import Address Table
              10018064 Import Name Table
                     0 time date stamp
                     0 Index of first forwarder reference

                 4A5 SetUnhandledExceptionFilter
                 4D3 UnhandledExceptionFilter
                 165 FreeResource
                 29C GetUserDefaultLangID
                 354 LockResource
                 341 LoadResource
                 14E FindResourceW
                 1C0 GetCurrentProcess

                       o
                       o
                       o

                 245 GetProcAddress
                 54D lstrlenA
                 3B1 RaiseException
                 367 MultiByteToWideChar
                 300 IsDebuggerPresent
                 511 WideCharToMultiByte
                 162 FreeLibrary
                 2E9 InterlockedCompareExchange
                 4B2 Sleep
                 2EC InterlockedExchange
                  CA DecodePointer
                  EA EncodePointer

    USER32.dll
              1001837C Import Address Table
              100181C0 Import Name Table
                     0 time date stamp
                     0 Index of first forwarder reference

                 333 wsprintfW
```

Figure 14-18. *Using dumpbin to list loading dependencies*

Dependency Walker

The Dependency Walker (a.k.a. depends.exe, see www.dependencywalker.com/) is the utility capable of tracking down the dependency chain of loaded dynamic libraries (Figure 14-19). It is not only capable of analyzing the binary file (in which case it parallels the Linux ldd utility), but it also can perform the runtime analyses in which it can detect and report the runtime dynamic loading. It was originally developed by Steve Miller, and was part of the Visual Studio suite of tools up until the VS2005 version.

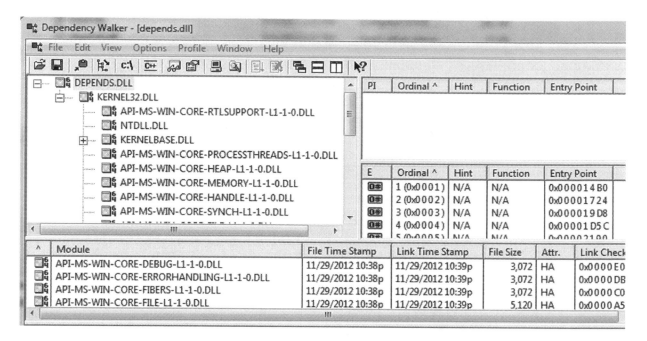

Figure 14-19. *Using the Dependency Walker utility*

Index

Get the eBook for only $10!

Now you can take the weightless companion with you anywhere, anytime. Your purchase of this book entitles you to 3 electronic versions for only $10.

This Apress title will prove so indispensible that you'll want to carry it with you everywhere, which is why we are offering the eBook in 3 formats for only $10 if you have already purchased the print book.

Convenient and fully searchable, the PDF version enables you to easily find and copy code—or perform examples by quickly toggling between instructions and applications. The MOBI format is ideal for your Kindle, while the ePUB can be utilized on a variety of mobile devices.

Go to www.apress.com/promo/tendollars to purchase your companion eBook.